CIRIA C519 London, 1999

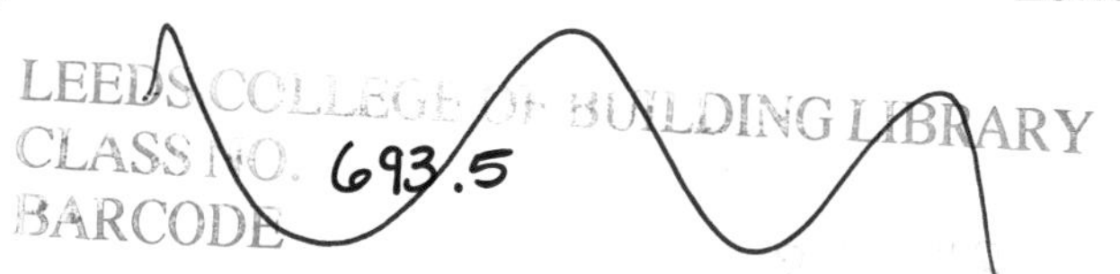

LEEDS COLLEGE OF BUILDING LIBRARY
CLASS NO. 693.5
BARCODE

Action in the case of nonconformity of [illegible]

P R Ainsworth BSc CEng FIStructE MICE MIHKE FCIArb

C J Hopkins BSc PhD

LEEDS COLLEGE OF BUILDING
WITHDRAWN FROM STOCK

sharing knowledge ▪ building best practice

6 Storey's Gate, Westminster, London SW1P 3AU
TELEPHONE 0207 222 8891 FAX 0207 222 1708
EMAIL enquiries@ciria.org.uk
WEBSITE www.ciria.org.uk

Summary

The purpose of this report is to assist those involved in the construction process, when faced with nonconformities, to arrive at appropriate solutions, in the shortest possible time and with minimum conflict.

The report sets out basic principles for dealing with nonconformities. These provide a rational framework against which nonconformities can be verified/investigated, their significance can be assessed, appropriate remedies can be considered and recurrence can be avoided.

Six key areas in which nonconformities frequently occur have been targeted in this report. These are: cover, inherent concrete quality, cracking, surface finishes, honeycombing and surface damage, and dimensional nonconformity. The report incorporates *inter alia:* flow charts setting out the decision-making processes for dealing with the six types of nonconformity; lists of prompts for consideration when carrying out investigations or when selecting appropriate remedies; and typical examples of nonconformities. General advice on specifying to reduce nonconformities and avoid disputes is also provided.

HEALTH AND SAFETY

Construction activities, particularly on construction sites, have significant health and safety implications. These can be the result of the activities themselves, or can arise form the nature of the materials and chemicals used in construction. This report does not endeavour to give comprehensive coverage of the health and safety issues relevant to the subjects it covers. Readers should consult other specific guidance relating to health and safety in construction.

Action in the case of nonconformity of concrete structures

Ainsworth P R and Hopkins C J

Construction Industry Research and Information Association

Publication C519 © CIRIA 1999 ISBN 0 86017 519 7

Keywords

Concrete, nonconformity, cover, inherent concrete quality, cracking, surface finish, honeycombing, surface damage, dimensional nonconformity, specification, contractual disputes.

Reader interest	Classification	
Designers, specifiers, constructors and supervisors involved in concrete construction. Concrete and component material suppliers. Mediators, conciliators, adjudicators.	Availability	Unrestricted
	Contents	Guidance to good management and construction practice
	Status	Committee-guided
	User	Professionals concerned with concrete construction

Published by CIRIA, 6 Storey's Gate, Westminster, London SW1P 3AU. All rights reserved. No part of this publication may be reproduced or transmitted in any form or by any means, including photocopying and recording, without the written permission of the copyright holder, application for which should be addressed to the publisher. Such written permission must also be obtained before any part of this publication is stored in a retrieval system of any nature.

Acknowledgements

The project leading to this report is part of a collaborative programme between CIRIA and The Concrete Society entitled "Concrete Techniques", and was carried out under contract by Wardell Armstrong – AMEC Civil Engineering Ltd – Knight & Sons (Solicitors).

Research team

P R Ainsworth	Wardell Armstrong
C J Hopkins	Wardell Armstrong
D P McNicholl	Wardell Armstrong
J B Wilkinson	Wardell Armstrong (architectural adviser)
D A Billington	AMEC Civil Engineering Ltd
E Hall	AMEC Civil Engineering Ltd
S Belshaw	Knight & Sons (legal adviser)

Steering group

The project was carried out and the report prepared under the guidance of the following steering group:

Mr R A McClelland (chairman)	Alfred McAlpine Construction Ltd
Dr P Bamford	Taywood Engineering Ltd
Mr B Brown	Ready Mix Concrete (UK) Ltd
Mr R Cather	Arup Research & Development
Dr C Clear	Civil and Marine Slag Cement Ltd
Mr C Ecob	Mott MacDonald, Special Services Division
Mr J Frearson	Rugby Cement (formerly Messrs Sandbergs)
Mr G Hammersley	Luton University
Dr T Harrison	Quarry Products Association
Mr D Ives	Highways Agency
Dr E Kay	Sir William Halcrow and Partners
Mr L H McCurrich	FOSROC International Technology
Dr B Marsh	Arup Research & Development (formerly BRE)
Mr M Mesham	Gibb Ltd
Dr J B Newman	Imperial College of Science and Technology
Mr P Owens	Quality Ash Association
Mr P F Pallett	Consultant
Mr R Roberts	Concrete Advisory Service
Mr G P Tilley	Gifford and Partners
Mr P Titman	Edmund Nuttall
Mr C Turton	Design Group Partnership
Mr M Walker	The Concrete Society
Mr D Wilson	CIS Construction Products
Mr P Woodhead	Department of the Environment
Dr B W Staynes	CIRIA research manager

The project was funded by the Department of the Environment, Transport and the Regions, CIRIA and The Concrete Society. CIRIA, The Concrete Society and the authors are grateful for the help given in this project by the many individuals who provided case reviews, representatives of the concrete repair industry, and the contributors to the CIRIA workshop held at Imperial College, London in October 1996.

Contents

FIGURES

TABLES

LISTS

Abbreviations

BCA	British Cement Association
CIB	Comité Euro-International du Béton
EDM	electro-optic distance measurement
ggbs	ground granulated blast-furnace slag
grp	glassfibre reinforced plastics
NDT	non-destructive testing
opc	ordinary Portland cement
pfa	pulverised fuel ash
RAM	rapid analysis machine
SRPC	sulphate-resisting Portland cement
USPV	ultrasonic pulse velocity

1 Introduction

1.1 PURPOSE

In general, the essence of resolving disputes over nonconformity, as proposed in this report, is not to produce legal winners and losers. Rather, as advocated by Latham[1], it is to endeavour to achieve win-win situations, ie to arrive as quickly as possible at a pragmatic, commercial solution which is acceptable to all of the parties concerned.

Concrete is an excellent construction material with innumerable applications. Over the past century it has become, in its various forms, the most widely used material in the construction industry.

Realising all of its potential benefits such as strength, durability and the capacity to be moulded into complex shapes is not, however, a simple process particularly when economy and speed of construction are also important factors. Notwithstanding the construction industry's best efforts, problems arising from nonconformities often occur.

The purpose of this report is to assist those involved in the construction process, when they are confronted with nonconformities, to arrive at appropriate solutions in the shortest possible time and with minimum conflict. The report should be of assistance to all parties seeking to resolve problems and disputes arising from nonconformities; not only employers, contractors, engineers and architects, but also third parties charged with resolving such disputes (mediators, conciliators, adjudicators, etc).

For this report, "nonconformity" means that which does not conform to the specification, and includes cases where it has been found that conformity with the specification is not possible. "Specification" is taken to mean any technical standard to which a contractor is required to conform and therefore includes drawings and ancillary documents.

Health and safety considerations were not included in the study leading to the report.

1.2 BACKGROUND

As construction techniques have become more sophisticated and the requirements of contract programmes have become more demanding, nonconformities and the attendant problems they create have become increasingly commonplace and difficult to resolve. Such problems are rarely considered until they arise (or, worse, until they can no longer be ignored), and then delays and costs begin to escalate as the fundamental questions of cause and effect are investigated. The pressures underlying these investigations are often such that they lead the parties involved to take up entrenched, self-protecting positions rather than working together to find equitable, practical and cost-effective solutions.

A review[2] has suggested that the primary causes of nonconformities lie in design and construction in roughly equal proportions, and that the costs of carrying out remedial work to defects in concrete construction in the UK are extremely high. There is no reliable data on the total number of days lost in remedying defects and in resolving the disputes to which they frequently give rise, but the total is thought to be large.

It is believed that both delays and costs can be significantly reduced if, in the spirit advocated by the Latham Report[1], the emphasis can be shifted to resolving problems as quickly and cost-effectively as possible, in the mutual interests of all concerned, rather than allowing confrontational situations to develop.

Against this background, CIRIA and The Concrete Society initiated this research project which is intended to assist in rational, objective decision-making. The project has been directed by a steering group composed of representatives from the various sectors of the concrete construction industry. The research team included individuals from various backgrounds including civil and structural engineers, materials specialists, an architect and a contract specialist. In compiling the report, information, advice and comment have been sought from a wide spectrum of construction industry representatives. The report is written from an independent standpoint. Given its broad basis it is believed that the advice presented will be acceptable to the construction industry as a whole.

1.3 OBJECTIVES

The aims of this report are to reduce controversy, delays and expense by:

- providing a rational framework within which cases of nonconformity can be investigated
- identifying the key considerations in specific cases
- where possible, providing quantitative means by which the significance of the nonconformity can be assessed
- recommending appropriate remedial actions
- encouraging those involved to learn from the experience so that recurrence can be avoided.

1.4 SCOPE

In order to keep the present project to manageable proportions, the following key areas in which nonconformities frequently occur have been targeted as its main focus:

- cover (Section 3)
- inherent concrete quality (Section 4)
- cracking (Section 5)
- surface finish (Section 6)
- honeycombing and surface damage (Section 7)
- dimensional nonconformity (Section 8).

It is suggested, however, that the general principles set out in this publication for dealing with nonconformities (Section 2) in these selected areas can be extended to other areas of concrete construction and, indeed, other construction materials.

Section 9 contains outline guidance on incorporating into contract specifications some of the principles discussed in Sections 3–8, with a view to minimising the incidence of nonconformities and avoiding delays and disputes when nonconformities do occur.

1.5 RESEARCH METHODOLOGY

The advice contained here has been formulated based on four main sources of information.

1. A review of the relevant published material has been carried out.
2. A series of case reviews has been drawn from discussions with a broad cross-section of practitioners involved in the various aspects of the construction industry. A total of 154 case reviews, obtained from 42 sources, has been collated and reviewed.
3. Discussions have been held with the concrete repair industry on techniques available for remedying some of the problems commonly arising from nonconformities.

4. A special workshop was held at Imperial College, London, at which views and experiences relating to dealing with nonconformity were exchanged.

1.6 STRUCTURE OF THIS REPORT

Section 2 of this publication sets out in general terms the issues to be considered in deciding how best to address a nonconformity. While many of the points made might seem obvious, case reviews suggest that some of these considerations can be overlooked and that unnecessarily hasty decisions are sometimes made. The rationale of the decision-making process is set out in order to act as an *aide memoire* to those dealing with a nonconformity.

In Sections 3–8 the basic principles set out in Section 2 are applied to the six generic types of nonconformity that are the subject of the report; Section 9 deals with specification issues.

2 General approach to dealing with nonconformities

2.1 THE CAUSES OF NONCONFORMITIES

Although a review of site practices[2] has suggested that the numbers of nonconformities arising from design shortcomings (ie unrealistic specifications or unworkable details) are roughly comparable with the numbers arising from poor workmanship or inadequacies in site management, the exact proportions in which these various factors contribute to the incidence of nonconformities cannot be reliably established. It is clear, however, that deficiencies in both design and construction can give rise to nonconformities, and that many nonconformities arise as a consequence of failings (or misunderstandings) in both.

2.2 THE RELEVANCE OF QUALITY ASSURANCE SYSTEMS

Increasingly, contractors are operating quality assurance (QA) schemes on sites, and checks that the structure conforms to the specification form part of such systems. The view that implementing a QA system will eliminate nonconformities is, however, erroneous. QA systems should be capable reducing the number of non-conformities and provide a means of detecting, relatively quickly, a nonconformity when it has occurred. They should also assist in tracing the origins of a nonconformity.

2.3 COMMERCIAL AND CONTRACTUAL CONSIDERATIONS

As discussed in Section 1, the objective of this report is to assist in achieving win-win situations, ie to assist in arriving as quickly as possible at pragmatic, commercial solutions which are acceptable to all parties. Once a nonconformity has been detected, actions have to be taken. There is a need to act quickly and, more importantly, to act rationally in order to rectify the situation. Any actions seeking satisfactorily to resolve a nonconformity cannot, however, disregard commercial and contractual concerns.

Specifications are usually incorporated into contracts, such that nonconformity with the contract specification constitutes a breach of contract. When a breach occurs, the strict contractual position is that the contractor must remedy it, provided it was physically and legally possible to comply with the terms of the contract in the first place.

Provided conformity was possible, the employer is entitled to be placed in the position in which they would have been, had the contract been satisfactorily performed. Strict application of contract conditions (such as the ICE Form of Contract) might require the removal and replacement of the noncompliant work or, as a minimum, will require the nonconforming element to be brought up to a standard equivalent to what was specified, such that its value and usefulness of the employer is in no way diminished.

There may be occasions, however, where an employer chooses to waive the contractual rights (usually in the interests of expediency) and accept a product which does not comply with the specification. They are, of course, entitled to do this and on such occasions may well look to their technical advisers for guidance on the implications of such a decision. However, it may be the case that technical advisers, appointed as engineer or architect under a contract, cannot make such a decision themselves on behalf of the employer without their consent. Care is therefore required in dealing with the acceptance of nonconforming elements.

The employer has a duty under common law to mitigate losses arising from nonconformity. It is required to act reasonably, and not knowingly to allow the costs of resolving a nonconformity to escalate unnecessarily.

2.4 THE FUNDAMENTAL ISSUES

If a nonconformity occurs, some fundamental questions have to be addressed.

1. Was conformity practicable?
2. What options are available for remedying the nonconformity?
3. What is the employer prepared to accept?

To answer the first question, an appraisal of the specification with regard to buildability is needed. If conformity was not possible, the design must be reviewed.

For the second question, an investigation into the nature and the extent of the nonconformity and an assessment of its significance is required. Only when this has been evaluated can options for remedying the situation be considered.

Answering the third question necessitates an understanding of the consequences of implementing any remedial actions.

2.5 THE BASIC APPROACH

When a nonconformity occurs or is suspected to have occurred, there is a series of basic steps to be followed:

- confirm or otherwise that a nonconformity has occurred
- prioritise future actions in terms of their practical effects and (if relevant) safety implications
- establish, if possible, the cause(s) of the nonconformity
- determine the extent of the nonconformity
- evaluate the significance of the nonconformity in the specific situation
- assess possible remedial options
- decide on appropriate remedial actions
- execute the necessary remedial action
- agree action to prevent or minimise the recurrence of nonconformities of the same type
- provide feedback via management systems so that improved specifications and practices can be developed.

A flow chart showing the interrelationship of these various steps is given in Figure 2.1.

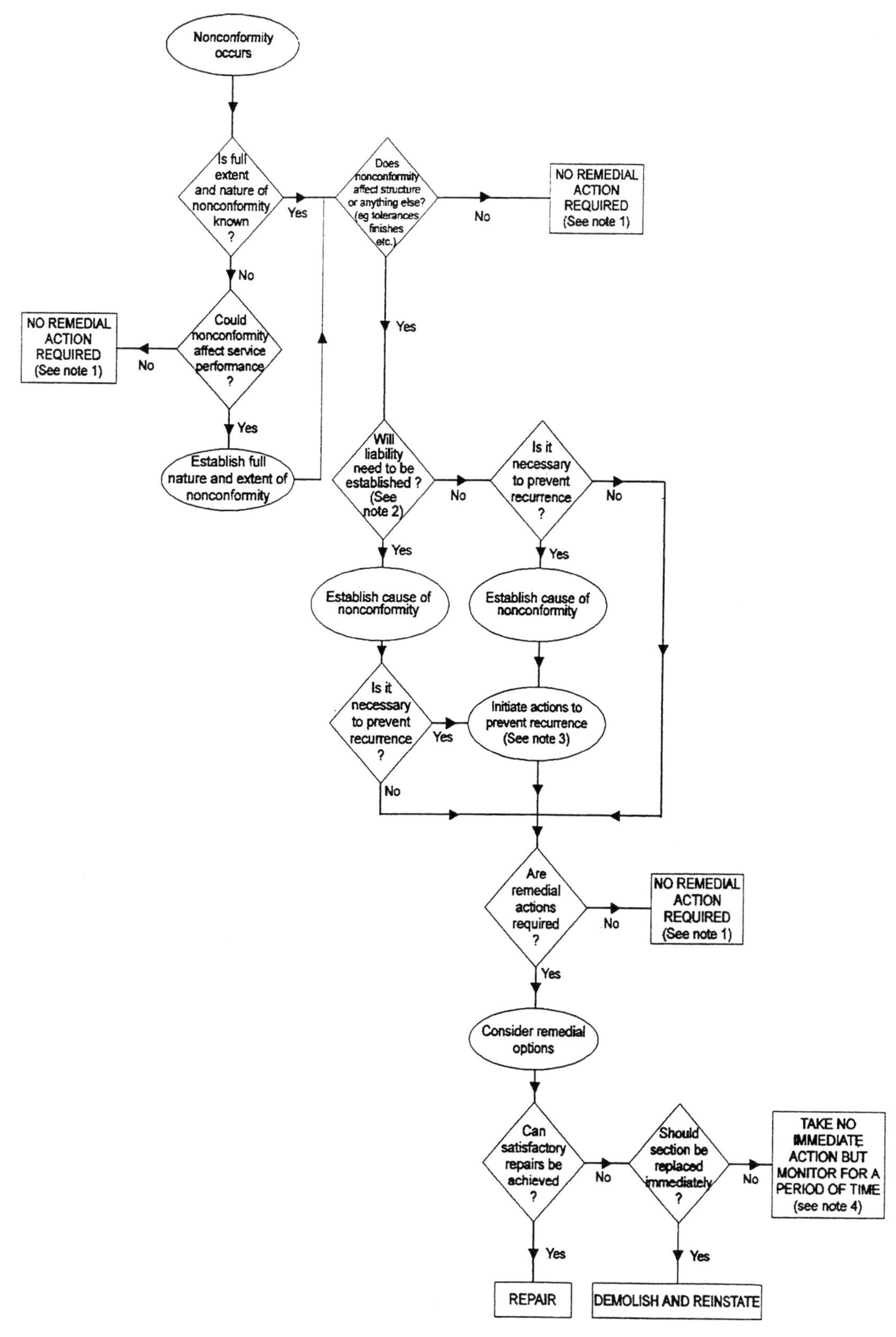

NOTES:

1. Refers to physical remedies only; action under the contract (issue of variation, financial adjustment, etc) may be warranted.
2. Liability need not necessarily be established at this stage; the question to be addressed is only whether it will have to be established.
3. This action does not have to be a precursor to subsequent actions in the flow chart, it can be undertaken in parallel.
4. The contractual implications of pursuing this option will need to be considered.

Figure 2.1 *Flow chart: general approach*

The steps can be classified into four processes, which are considered in general terms in the remainder of this chapter and more specifically in Sections 3–8.

1. **Verification/investigation.**
2. **Establishing the significance of the nonconformity.**
3. **Determination of appropriate remedy.**
4. **Prevention of recurrence.**

2.5.1 Verification/investigation

The essential approach to establishing and executing a verification/investigation programme is set down below.

1. **Decide if an investigation is warranted.** It may be obvious from the outset that certain minor nonconformities have no significance with regard to the performance of the structure and therefore need not be subjected to expensive and time-consuming investigations.
2. **Consider the effects, in terms of delay and disruption, of carrying out an investigation.**
3. **Agree, if possible, who is to bear the cost of the investigation (including any design checks).** This may, of course, be on the basis that the costs will be borne by one party in the event of one outcome and by another party in the event of a different outcome.
4. **Ensure, if possible, that all parties concerned agree with the aims and methodology of the investigation**. If such agreement cannot be achieved it may be necessary for one party to proceed unilaterally, but this approach should be avoided if at all possible.
5. **If possible, establish before the investigation commences what courses of action are to be pursued in case of the various possible outcomes.** Acceptance criteria for the investigation results, and any further actions to be taken in the event of unsatisfactory results, should be agreed before the investigation commences.
6. **Choose an investigation method which is appropriate for the nonconformity in question**. That is to say, ensure that it either measures directly the parameter specified or, where different, the property actually required. (Appropriate investigation methods may be prescribed in the specification.)
7. **Select an investigation method which gives a precision of answer commensurate with the tolerance given in the specification.**
8. **Where possible, choose investigation methods that are simple, economical, can be implemented quickly and give rapid results.** Investigation may not mean testing; visual inspection only may be necessary, or a review of records may yield sufficient information.
9. **Decide whether the objective of the investigation is to establish only the degree and extent of the nonconformity or also its cause.** The latter may not be necessary for deciding upon an appropriate remedy, but it probably will be necessary if responsibility is to be determined, and it certainly will be if recurrence is to be prevented.
10. **Ensure that the investigation will of itself be conclusive**. In some cases, the conclusion may be that more tests of the same type are required, more sophisticated tests are required or monitoring is necessary either in the short or long term. This situation should be avoided, however, unless absolutely unavoidable. Where possible, the investigation should lead directly to the decision on appropriate remedial action.

11. **Try to ensure that the amount of work at risk, pending the outcome of the investigation, is kept to a minimum.** It is clearly preferable to avoid casting further concrete on top of suspect concrete until the implications of any shortcomings have been evaluated. If further concreting is unavoidable, the risks associated with proceeding will need to be considered. Alternatively, if work is suspended pending the results of the investigations, the practical and cost implications of consequent delays will need to be considered.

12. **Where the chosen investigative technique is particularly sophisticated or complex, use specialists where appropriate.** Poorly conducted investigations may give inconclusive answers which serve to exacerbate rather than help to resolve the situation. Whenever possible, there should be prior agreement as to who is to undertake the investigation.

In considering testing schedules, the following general points should be borne in mind.

1. There is sometimes merit in correlating a large number of relatively inexpensive tests with a small number of more expensive tests.

2. Test locations should be chosen carefully, both in terms of their representativeness of the concrete work at issue and in terms of their effect on the structure.

3. Extracting samples for testing or executing in-situ testing is usually destructive. Many so-called non-destructive tests do in fact damage the surface of the concrete.

4. Nonconformities are usually detected in young concrete, during or soon after the completion of construction. The age of the concrete at the time of testing may influence the results, and could also influence the choice of test method.

5. Account should be taken of the reliability of the investigation technique and of the level of uncertainty associated with testing[3].

2.5.2 Establishing the significance of nonconformities

The significance of any nonconformity will need to be assessed with regard to:

- structural integrity
- suitability of the structure for its intended use
- durability
- aesthetics
- the effect on subsequent operations.

The assessment of the significance of a nonconformity will clearly be largely a matter of engineering judgement, but the following questions are likely to need to be addressed.

1. What was the intention of the original design? How is that intention affected by the nonconformity?

2. What was the intention of the specification? Was the parameter specified what was actually required or was it simply a means of achieving what was required?

3. What is the intended purpose of the structure or part of the structure in question? how and to what extent is that purpose affected by the nonconformity?

4. What is the location of the nonconformity? Is it critical?

5. What is the extent of the nonconformity?

6. What is the relevance of the specification to the nonconformity in question?

2.5.3 Determination of appropriate remedies

Once the nature, extent and significance of the nonconformity have been determined, remedial options can be assessed. Minor nonconformities may require no action; severe cases may demand the removal and replacement of the cast sections. Many nonconformities will warrant action somewhere between these two extremes.

Remedial options need to be carefully evaluated and the following questions addressed.

1. Does the proposed remedy place the employer in exactly the position he would have been had the contract been satisfactorily performed, ie had the nonconformity not occurred? In reality, it rarely does unless the solution is to demolish and reconstruct, and even then there may well be adverse implications for the employer.
2. Does the proposed remedy place the employer in a position equivalent to the one in which he would have been had the nonconformity not occurred? If not, does the employer concur with the proposed course of action; does he expect somehow to be compensated?
3. Is the claimed effectiveness of the proposed remedy guaranteed? If so, how and by whom? What are the risks?
4. What are the practical implications of implementing a proposed remedy, in terms of potential effects on surrounding structures, disruption, delay, hazard, etc?
5. What are the future maintenance implications of the proposed remedy? Definite guidance should be provided for the future owner or operator of the structure.
6. What are the aesthetic implications of the proposed remedy?
7. What is the cost of the proposed remedy – both immediate and long-term (with regard, for instance, to ongoing maintenance)? How and by whom is it to be borne?

2.5.4 Prevention of recurrence

It is not always possible to establish the causes of nonconformity. If, however, the causes can be established, recurrence can usually be avoided. The cause may lie with:

- design
- detailing
- specification
- materials
- construction techniques
- workmanship
- lack of communication
- combinations of these.

Several questions need to be addressed.

1. Was the nonconformity an inevitable occurrence? For instance could what was specified be achieved, or was the problem inherent in the nature of the structure?
2. Was the nonconformity the consequence of the acts or omissions of an individual or group of individuals? Such acts or omissions could be careless, accidental or wilful.
3. Was the nonconformity the consequence of the method of construction?

Positive answers to these questions lead to the following considerations.

1. The practicality of the original design/specification needs to be reviewed. Does the structure need to take the exact form designed or is there a different but practicable and equally effective alternative? Was it really necessary to achieve what was specified; was it possible to achieve what was specified? Did the contractor pass any comment upon buildability at tender stage?
2. Careless acts or omissions may be averted through improved QA procedures; wilful acts may be averted through closer supervision or changes in personnel; accidental acts or omission may be averted through review of QA or working procedures.
3. If the nonconformity has arisen as a consequence of the method(s) of construction, those methods need to be reviewed and where appropriate amended.

In respect of 2 and 3 above, it might be that a nonconformity has occurred primarily as a consequence of poor workmanship or inappropriate construction methods, but the situation has been exacerbated by poor design or specification. That is to say, what was specified was unnecessarily difficult to achieve and the nonconformity was waiting to happen. In such cases both the design and the construction methods need to be reviewed.

It is recognised in such cases, however, that practical considerations may to some extent conflict with contractual and commercial ones. If the fault for a nonconformity can be seen to lie with a contractor, the engineer may be unwilling to acknowledge, or even consider, the possibility that the design or specification might also have had an influence. It is suggested, nevertheless, that the primary objective of both parties should be to ensure that the recurrence of problems is minimised, and that if both sides take a professional and reasonable position this can be achieved equitably and without conflict.

2.5.5 Pre-empting disputes

The occurrence of nonconformities can cause disputes and delays. Many of the disputes associated with more commonly occurring types of nonconformity can be avoided, however, by setting out in the specification procedures to be followed in the event of such a nonconformity. This subject is dealt with in Section 9.

2.5.6 Monitoring

Where long-term monitoring is identified as the appropriate course of action for dealing with a nonconformity, it is important that the parties involved at the time of construction define what is to be monitored and what action is to be taken in the event of specific observations, and that they provide definitive guidance for the future owner or operator of the affected structure.

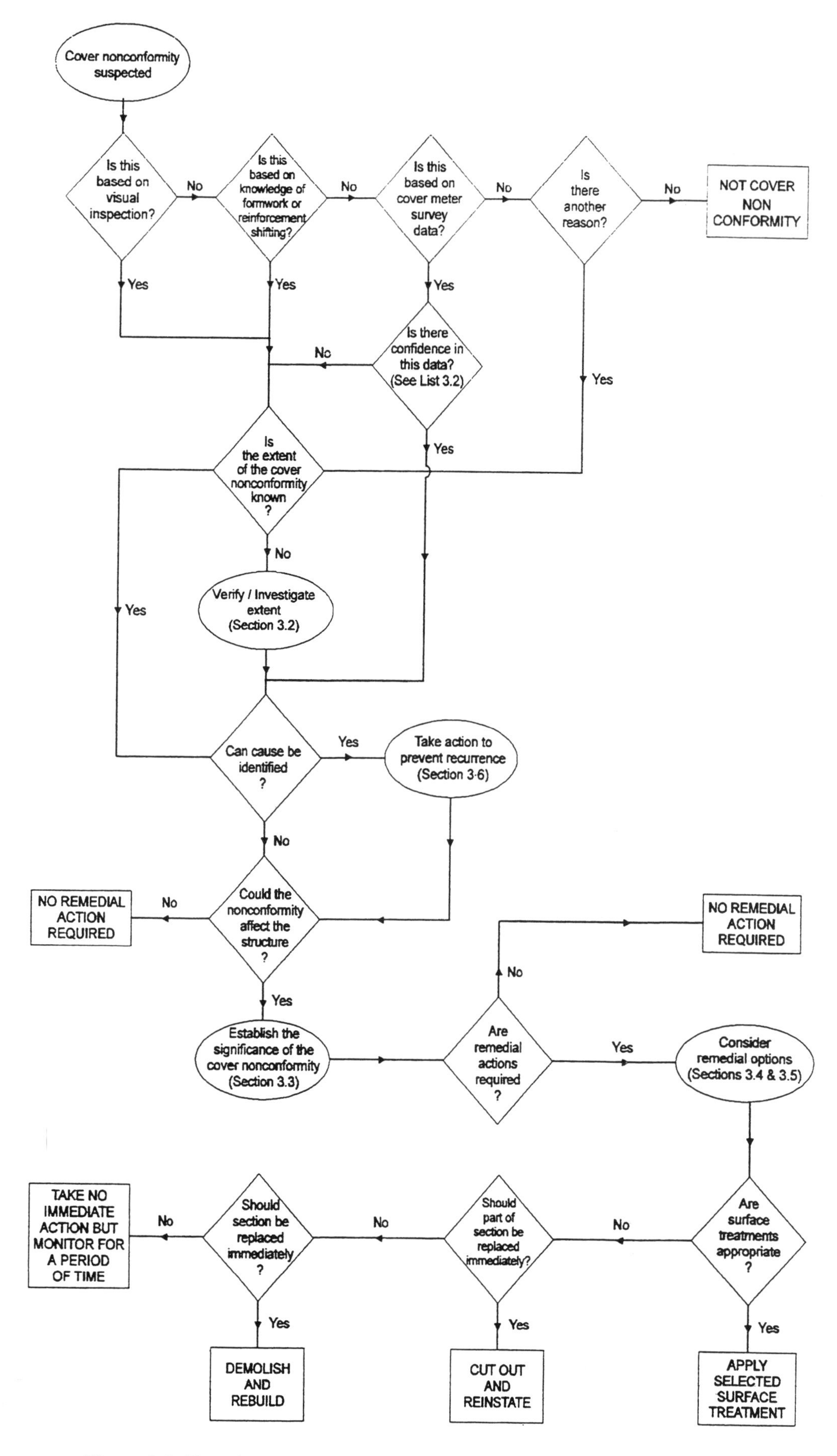

Figure 3.1 *Flow chart: cover*

3 Cover

3.1 INTRODUCTION

The decision-making framework for dealing with a suspected cover nonconformity is set out in the flow chart in Figure 3.1.

As described in Section 2, there are four key processes.

1. Verification/investigation.
2. Establishing the significance.
3. Determination of appropriate remedy.
4. Preventing recurrence.

This section of the report identifies the specific issues that need to be addressed when applying these four processes to dealing with a suspected cover nonconformity.

3.2 VERIFICATION/INVESTIGATION

3.2.1 Available techniques

If a cover nonconformity is suspected there are several techniques available to verify that it has occurred and to investigate its extent.

1. Detailed visual inspections (Section 3.2.2).
2. Covermeter surveys (Section 3.2.3).
3. Other specialist technique: radar (Section 3.2.4).

Assessments may also need to be made of the concrete quality to help in establishing the significance of any cover nonconformity.

3.2.2 Detailed visual inspection

Visual inspections of the structure are often undertaken as a preliminary exercise. They should, nevertheless, be carried out carefully. Signs of exposed reinforcement should be noted, as should any surface damage or cracking. It is sometimes possible to see very low cover reflected in the surface of concrete as a colour change due to local differences in the degree of compaction. Evidence of corrosion on the concrete surface should be treated with caution, however, unless the reinforcement is actually visible. Even where negligible cover has been provided, it will often be the case that corrosion will not commence for several months, and so where brownish rust stains are observed they may not be indicative of insufficient cover. Rather, they may be due to the corrosion of loose nails or tying wire on the concrete surface, or due to the presence of pyrites in the aggregate, or associated with rusty starter bars protruding from the recently cast section.

A list of prompts for consideration when carrying out visual inspections to verify a cover nonconformity is given in List 3.1.

Prompts for consideration when carrying out a visual inspection to verify a cover nonconformity
1. Are any rust stains visible?
2. Are there are loose nails or pieces of tying wire embedded in the concrete surface which might be causing rust staining?
3. Is any reinforcement exposed on the surface of the concrete?
4. Is any reinforcement visibly out of position at construction joints?
5. Is any reinforcement visibly out of position at stop-ends?
6. Is any reinforcement visibly out of position at column or wall kickers?
7. Does the concrete appear to be well compacted?
8. Is there any surface colour change die to local differences in compaction as a result of low cover?
9. Is there any spalling of concrete or evidence of other surface damage?
10. Are there any cracks present?
11. What is the size of the cracks?
12. Do any of the cracks lie along the line of reinforcement?

List 3.1 *Prompts for consideration when carrying out a visual inspection to verify a cover nonconformity*

3.2.3 Covermeter surveys

Undertaking a covermeter survey is the most common means of investigating cover depths. Reliance should not be placed on data acquired from covermeters that have not been calibrated for the specific conditions pertaining to the particular investigation.

Description of technique

Electromagnetic covermeters operate by generating an electromagnetic field which is distorted by ferrous objects. Local changes in the electromagnetic field strength are detected, and provide a means of establishing the position (and size) of reinforcing bars.

Application and limitations

Properly used, covermeters can provide a rapid, non-destructive means of checking the location of reinforcement and the cover depth[4,5]. They are best suited to lightly reinforced elements. Modern instruments are battery-operated, relatively light, portable and easy to handle. Sophisticated meters can provide information on the type and size of reinforcing bars, as well as their position. Basic covermeter models operate over two cover ranges: 0–40 mm and 40–100 mm. More sophisticated models are capable of measuring cover depths of up to 200 mm.

Epoxy-coated and galvanised bars can be detected with covermeters but detecting some types of stainless-steel reinforcing bars can be difficult. Bars made from non-ferrous materials cannot be detected by covermeters.

Many factors can adversely influence the output data from a covermeter, leading to erroneous cover depth evaluations, so considerable care is needed when conducting a covermeter survey. In most cases it will be essential that covermeter readings are checked and calibrated, by drilling or by breaking out in localised areas to establish true cover depths for comparison with the meter readings. It is helpful to refer to a set of reinforcement drawings when interpreting the results of a covermeter survey. A list of prompts for consideration when carrying out covermeter surveys is given in List 3.2.

Prompts for consideration when carrying out an electromagnetic covermeter survey
1. Has the scope of the covermeter survey been agreed at the outset?
2. Is the survey to be carried out by a UKAS laboratory or other accredited organisation?
3. Has the survey been carried out in accordance with BS 1881[5]?
4. Are personnel experienced in conducting cover surveys?
5. Are the batteries fully charged and spare batteries available?
6. Have the manufacturer/s instructions been followed and the correct opening conditions used for the survey?
7. Is the covermeter using eddy current or magnetic induction? (The former may require special calibration for steel type and concrete conductivity.)
8. Has the instrument been zeroed correctly, away from the structure and other metals (eg scaffolding, steel pipes, window fixings)?
9. When was the instrument calibrated in the laboratory?
10. Has the instrument been calibrated on site?
11. Have frequent on-site calibration checks been made during the survey?
12. Have the cover readings been cross checked with true *in situ* cover values by drilling or breaking out concrete?
13. Is the concrete surface damp?
14. Does the concrete have a rough surface or a special concrete finish that might influence the readings?
15. Does the concrete contain aggregates or cement composites which can influence the readings?
16. What is the reinforcement configuration, bar type and bar diameter?
17. Are the bars made of non-ferrous material (since the cover to these types of bars cannot be measured using a covermeter?
18. Are the bars straight or curved?
19. Are the bars parallel to the concrete surface?
20. Are any loose nails, pieces of tying wire, etc influencing the readings?
21. Are any stray magnetic readings affecting the readings?
22. Are any bundles, laps or couplers affecting the measured cover value?

List 3.2 *Prompts for consideration when carrying out an electromagnetic covermeter survey*

Reliability of covermeter surveys

Research shows that the on-site accuracy of cover evaluation achievable by experienced covermeter operators is ±15 per cent (or ±5 mm, whichever is greater)[5,6]. Greater deviations can be anticipated for inexperienced users.

Reinforcement configurations can influence the accuracy of cover measurement. As the complexity increases, it becomes more difficult to identify individual bars, which can lead to erroneous cover measurements. Figure 3.2 illustrates increased degrees of difficulty. For closely spaced parallel bars, problems in locating individual bars usually arise when the spacing is less than twice the cover depth[7].

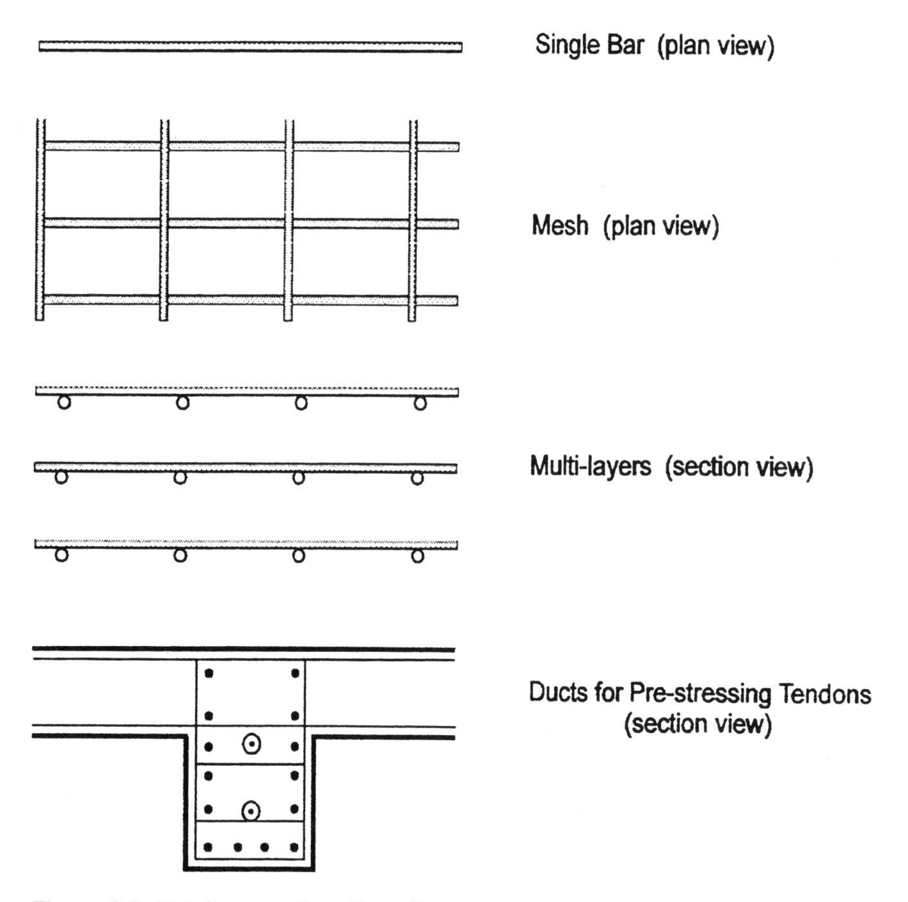

Figure 3.2 *Reinforcement configurations*

Other factors that can adversely affect the accuracy of covermeter readings include:

- certain aggregates that contain minerals with magnetic properties
- cement and cement composites (SRPC, pfa, ggbs, etc) that have variations in iron content
- loose nails or tying wire lying within the cover zone.

The size of the probe head can also affect the accuracy of the readings. For rough or undulating concrete surfaces the irregularities under the search head will affect the cover readings. Difficulties may also be encountered at the edges of sections and on soffits.

3.2.4 Other specialist technique: radar

Description of technique

This is a relatively new and rapidly developing technique in which a pulse of electromagnetic energy is directed into the concrete, and reflected signals are detected. The strength and the time delay associated with the reflected signal enable depths to reinforcement bars to be calculated, with the bars producing a characteristic hyperbolic reflection pattern. Surface penetrating radar equipment typically operates at a frequency of 1 GHz for cover depth investigations[8].

Application and limitations

Radar may be an appropriate investigation technique for cover depth if large areas of a structure need to be examined quickly. The required equipment, however, is less readily available and less mobile than a covermeter. Again, supplementary drilling or breaking out of concrete to check actual cover values is required. The operation of radar and the interpretation of scan patterns need to be undertaken by experienced radar users.

Reliability

Estimations of cover depth are unlikely to achieve an accuracy of better than ±5 mm[8], and errors of up to 45 per cent have been encountered for small cover depths[9]. Concrete moisture content also influences the signal velocity which, because newly constructed concrete has a relatively high moisture content, can lead to over-estimations of cover depth on relatively recently cast sections.

3.2.5 Assessment of concrete quality

For reasons which are discussed in Section 3.3.2, it will frequently be the case that assessments of concrete quality will have to be undertaken alongside assessments of cover. Guidance on the assessment of concrete quality is given in Section 4.

3.2.6 Extent of investigation required

When conducting an investigation to verify whether a cover nonconformity has arisen, consideration should be given to whether the nonconformity is localised or whether a more widespread investigation is warranted. Once a cover nonconformity has been verified, it is prudent to extend the investigation to examine:

- work undertaken by the same operators or at the same time
- structural members of similar type
- locations such as slab and beam edges, construction joints, beam/column junctions, etc which are known to be prone to incidences of low cover (see Section 3.6.4).

3.2.7 Tolerances on specifications

Tolerances permitted by some design codes[10-12,14] typically allow actual cover values (ie minimum cover) to be up to 5 mm less than the quoted cover (ie nominal cover), provided that the minimum cover achieved exceeds the maximum nominal aggregate size and exceeds the main bar size. Testing standards[5] allow for tolerances of ±5 mm (or ±15 per cent, whichever is greater) on cover measurements of less than 100 mm. The accuracy of measurement achievable on site, even by an experienced covermeter operator, is unlikely to be better than ±15 per cent.

Consideration should therefore be given to the possibility of cumulative tolerances. For instance, specified nominal cover may be 40 mm; a spot covermeter check may indicate a cover of 35 mm – in theory acceptable – but the true cover could be 30 mm. Conversely, the covermeter could indicate a cover of 30 mm, which would be classed as a nonconformity, when the actual cover could be 35 mm, which would conform.

Checks on cover depths must set out to demonstrate that the required minimum cover value has been achieved.

The purpose of specified tolerances on cover values in UK design codes is to allow for any variation in placing the reinforcement during construction. Numerous investigations examining the accuracy of reinforcement placing in practice[15-18] have acknowledged that the present values for specified tolerances in design codes are not always achieved in practice, even with good workmanship. The European design codes[131] take a different approach, in that a required minimum cover is identified to which is added a margin. For in-situ reinforced concrete construction, the margin is in the range 5–10 mm. The 5 mm value may be adopted if production control can guarantee this and if it is verified by quality control. A margin of 10 mm is the norm, however, and is considered a practical tolerance that can be achieved with good workmanship. The nominal cover (minimum plus selected margin) is shown on the drawings.

Note: the current UK National Application Document gives one tolerance value of 5 mm for both precast and in-situ concrete.

LEEDS COLLEGE OF BUILDING
LIBRARY

3.3 ESTABLISHING THE SIGNIFICANCE OF NONCONFORMITIES IN COVER

3.3.1 The purposes of cover

The significance of a cover nonconformity must be assessed in relation to the purposes of cover. Concrete cover to reinforcement serves several purposes:

- protecting the reinforcement from attack by aggressive agents in the external environment (ie a means of providing a durable structure)
- providing protection to the reinforcement in the event of fire
- facilitating composite structural action of the reinforcement and the surrounding concrete
- satisfying bond requirements
- serving as an architectural finish or as a base for applied finishes.

For external concretes the durability requirement is usually the governing factor. That is, if the required amount of cover is provided for purpose 1 above, it will usually follow that there is also sufficient cover for purposes 2–5. For internal concretes, structural or fire requirements may dominate.

3.3.2 Cover requirements and the quality of concrete

In current UK codes and standards, recommended cover values are associated with concrete grade (strength), water/cement ratio, cement content and exposure conditions. Thus, if the quality of the cast concrete can be shown to be better than specified, a reduction in cover depth may be permissible for the same exposure conditions.

The concept of trade-off between concrete quality and cover value is included in some design codes[11,12], as an aid in the design of structures.

In circumstances where a cover nonconformity has arisen but the in-situ concrete quality can be shown to be superior to that specified, a similar trade-off between concrete quality and specified cover value may be possible in order to resolve a cover nonconformity. The concept of trade-off can be demonstrated by reference to Table 3.3 in BS 8110: Part 1[10], which is reproduced here as Table 3.1.

Table 3.1 *Nominal cover to all reinforcement (including links) to meet durability requirements (reproduced from BS 8110: Part 1[10])*

Conditions of exposure	Nominal cover dimensions in millimetres				
Mild	25	20	20	20	20
Moderate	–	35	30	25	20
Severe	–	–	40	30	25
Very severe	–	–	50	40	30
Most severe	–	–	–	60	50
Abrasive	–	–	–	–	–
Maximum free water/cement ratio	0.65	0.60	0.55	0.50	0.45
Minimum cement content (kg/m^3)	275	300	325	350	400
Lowest grade of concrete	C30	C35	C40	C45	C50

If, for example, the exposure conditions are deemed severe and a C40 grade concrete is specified, a nominal cover of 40 mm is required. Based on permitted tolerances, the actual minimum cover to be achieved is 35 mm. If, during investigations, the actual cover is measured as 30 mm, a cover nonconformity has occurred. However, if the concrete in-situ is shown to be equivalent to a grade C45, then for the same exposure conditions, reference to Table 3.1 shows that a nominal cover of 30 mm (which means an actual minimum cover requirement of 25 mm) would be acceptable, provided it can be demonstrated that the minimum cement content and maximum water/cement ratios of the in-situ concrete are applicable to the higher grade concrete requirements. Based on this trade-off, the measured cover value is satisfactory and the nonconformity is resolved. Conversely, if lower in-situ concrete quality and/or more severe exposure conditions are encountered, similar considerations lead to the conclusion that the nonconformity may be more serious than might originally have been thought.

Similar trade-off arrangements to this example can be made using BS 5400[11], but it might not be appropriate to adopt this approach for structures designed to codes[13,14], which do not include trade-off relationships.

Before considering a trade-off, it is important to recognise that the nominal cover values quoted in design codes were originally based on a survey of concretes containing Portland Cement and normal weight aggregate with a maximum nominal size of 20 mm, although the quoted cover values are also now frequently used for composite cement concretes. Trade-off is not normally permitted when bond or fire performance criteria dominate the specified cover requirements.

3.3.3 Exposure conditions

A similar trade-off may be possible between exposure conditions and required cover depths, in situations where the actual exposure conditions are found to be less severe than those originally envisaged.

It is important, therefore, that reference be made to the specification to check the assumed exposure condition upon which the design has been based. If it can be shown that the actual exposure condition is less severe than assumed in the design, a reduction in permissible cover depth may be justified.

3.3.4 Extent of nonconformity

The extent of the cover nonconformity needs to be evaluated. Isolated instances of lack of cover could be deemed to be acceptable providing that the locations and their exposure to aggressive agents are not critical.

3.3.5 Use of composite cement

Concretes containing composite cement such as pfa and ggbs may, for the same strength grade, result in concretes with lower water/cement ratios, lower permeabilities and better sulphate and chloride resistance properties. It may, therefore, be appropriate to review the specified cover depth if pfa or ggbs have been incorporated into the concrete mix but not provided for in the original design. It may be appropriate to consult experts on the characteristics of the constituent materials in these circumstances.

3.3.6 Subsequent finishes

It is useful to consider whether or not the section is to be clad or rendered at a later stage. Provided that the nonconforming cover does not affect any fixings into the concrete, reduced cover may not be significant if the applied finish will provide protection to the concrete and the reinforcement.

If the structure is to be clad, particular care needs to be taken to ensure that the rate of carbonation in the concrete behind the cladding is not increased.

Consideration should be given to the life expectancy of the applied finish compared to that of the concrete. It may be necessary to make provisions for replacing the applied finish in the future.

3.3.7 Poor compaction of concrete

Evidence of poor compaction of concrete in the vicinity of a cover nonconformity suggests that the in-situ concrete quality may be below that normally expected. Lack of cover is, therefore, more likely to be critical in these situations, although this cannot normally be quantified.

3.3.8 Reliability of investigation technique

In order to establish the significance of a cover nonconformity account must be taken of the reliability of the investigation technique *per se* and of the level of uncertainty associated with testing[3].

3.4 DETERMINATION OF APPROPRIATE REMEDY: EXCESS COVER

Where excess cover has been established it may not be necessary to take any remedial action, given that there will almost certainly be a consequent improvement in the protection to the reinforcement. However, it may be necessary to remove the excess cover for structural, aesthetic or dimensional reasons.

Where excess cover gives rise to a reduction in the lever arm at which the reinforcement is working in the concrete, the options include:

- recheck the design to establish whether the reduction in lever arm can be tolerated
- strengthen the section (for example by means of plate bonding)
- relax the specification
- break out and recast.

It should be noted that the first option can range from a simple reworked design calculation to a comprehensive appraisal of the as-built structure using in-situ strengths and modified safety factors[50].

3.5 DETERMINATION OF APPROPRIATE REMEDY: INSUFFICIENT COVER

Where insufficient cover has been established, the remedial options are:

- take no action if the nonconformity is not significant,
- take no immediate action but monitor,
- provide additional protection to the reinforcement, and if necessary increase the structural capacity,
- relax the specification,
- cut-out and reinstate,
- demolish and rebuild.

These options are discussed in turn below.

3.5.1 Take no action

Where the in-situ concrete is of a higher quality than that specified, or the exposure conditions have been found to be less severe than was assumed in the specification, it may be that the shortfall in cover can be tolerated with no further action required.

3.5.2 Take no immediate action but monitor

In certain circumstances it may be an acceptable to delay any action, but this decision must be based on a thorough understanding of the possible consequences of low cover in the particular situation. It would normally be appropriate in such cases to plan for regular monitoring, so that any signs of factors which could lead to early deterioration (ie significant carbonation or chloride contamination) are detected and appropriate remedial measures initiated before any corrosion of the reinforcement occurs. The monitoring should normally be begun within the first five years of the life of the structure.

Pursuing this option carries risks. In the longer term, it may be more difficult and costly to restore a structure which has begun to deteriorate. The contractual implications of this option will therefore need to be considered.

3.5.3 Provide additional protection to reinforcement

Providing additional protection equivalent to (or in excess of) the shortfall in cover usually involves the use of cementious renders or coatings. Surface treatments protect the reinforcement by reducing the permeability of the concrete so that aggressive agents are inhibited or prevented from initiating corrosion. To do this, surface treatments must control the ingress of:

- water
- carbon dioxide
- chloride ions.

Additional fire protection may also be required.

Selection of appropriate surface treatment

Selecting a surface treatment gives rise to a number of considerations, ie:

- protection properties
- method of application
- maintenance requirements
- durability of coating
- life-expectancy of coating
- recoating requirements
- aesthetics
- costs.

Comparisons between different types of surface treatment and between different formulations of the same generic type are difficult, since there are no standard tests at present. Manufacturers frequently use different test methods and a variety of performance criteria to describe their products. Guidance on the use of surface treatments is available[19,20].

A list of prompts to assist in selection is presented in List 3.3. A summary of the types of surface treatment presently available which are suitable for reinstating cover is presented in Table 3.2. It should be stressed that this summary is by necessity simplified, and more detailed specialist advice may need to be sought.

Prompts for consideration when selecting type of surface treatment suitable for providing protection to reinforcement in the event of a cover nonconformity
1. Does the treatment provide adequate resistance to water ingress?
2. Does the treatment provide adequate carbon dioxide diffusion resistance?
3. Does the treatment provide adequate resistance to chloride ion ingress?
4. Does the treatment provide adequate resistance to water vapour transmission (ie does the treatment have breathability)?
5. Is the surface treatment sensitive to UV light?
6. Is the surface treatment flexible and does it have crack bridging abilities?
7. Is the abrasion resistance of the treatment adequate?
8. How easily can the treatment be applied to the concrete surface?
9. Does the surface treatment adhere to concrete?
10. Does the surface treatment increase structural capacity?
11. Are there special requirements for temperature or curing?
12. Will a multi-protection system be required?
13. Are the different components of any multi-protection system compatible?
14. Is there a possibility of shrinkage?
15. How is the treatment affected by weathering?
16. What is the service life expectancy of the surface treatment?
17. Can the proposed treatment be overcoated at a later date?
18. Will the adhesion of subsequent finishes be affected?
19. Is the surface treatment aesthetically acceptable, both now and after several years of exposure?
20. How much does the (initial and recurring) treatment cost?
21. Could the proposed treatment increase the rate or risk of corrosion (eg by trapping moisture, reducing adhesion, reducing freeze/thaw resistance, increasing the rate of carbonation?

List 3.3 *Prompts for consideration when selecting type of surface treatment suitable for providing protection to reinforcement in the event of a cover nonconformity*

Many surface treatments perform better if applied to a dry concrete surface. However, new concrete is relatively moist, has a high alkalinity and is still reacting chemically. Also, while concrete will generally have achieved 80 per cent of its strength after 28 days it will only have undergone some 15–40 per cent of its shrinkage after six months. In some circumstances it may be appropriate to delay applying the surface treatment for up to six months, particularly if the proposed treatment has poor flexibility properties.

Surface treatments have a finite life, and they can fail prematurely. Causes include poor surface preparation, loss of adhesion, blistering, weathering, thermal expansion effects and incompatibilities between layers.

Opting to apply surface treatments to remedy a cover nonconformity may entail a commitment to regularly recoating the concrete, both for continued protection and for aesthetic reasons. It is important to agree who will be responsible for such longer-term maintenance and, if necessary, to set aside funds to meet the future costs of these works.

Type of surface treatment	Single or multi-component	Maintenance requirements	Chloride resistance	Carbon dioxide resistance	Breathability	UV light resistance	Flexibility	Other considerations
Cementitious only mortar	Single or multi	Should last lifetime of structure	Limited	Limited	Not applicable	Not applicable	Not applicable	Cannot use at low temperatures; cannot apply in thin layers; long-term bond and shrinkage problems
Polymer modified cementitious mortar	Single or multi	Inspect should last lifetime of structure	Yes	Yes	Not applicable	Not applicable	Not applicable	Cannot use at low temperatures; shrinkage and bond better than cementitious-only mortar
Screed/ render	Single or multi	Should last lifetime of structure	Limited	Limited	Not applicable	Not applicable	Not applicable	Cannot use at low temperatures; poor adhesion; difficult to apply; good weathering resistance
Silane/ siloxane penetrants	Single	Reapply after 3–5 years	Yes	No	Not applicable	Good	Not applicable	Cannot use at low temperatures; concrete must be 14 days old before applying; hydrophobic; colourless; freeze/thaw resistance; satisfactory penetration depends on concrete permeability
Elastomeric coating	Single	Recoat after 10/15 years	Yes	Yes	Yes, but need to avoid water collecting behind coating	Yes	Yes	Cannot use at low temperatures; pigmented types available
Acrylic coating	Single or multi	Recoat after 10/15 years	Yes	Yes	Yes, but depends on formulation	Good	No, although newer acrylic formulations have better flexibility	Cannot use at low temperatures; emulsion or solvent types available; some have poor bond to damp surfaces; pigmented or clear types available
Epoxy coating	Single or multi	Should last lifetime of structure	Yes	Yes	Limited	Poor	Not as flexible as polyurethanes	Cannot use at low temperatures; gloss finish; overcoating causes bloom; some have poor bond to damp surfaces; minimal shrinkage
Polyurethane coating	Single or multi	Recoat after 10/15 years	Yes	Yes	Yes, but depends on formulation	Good	Yes	Cannot use at low temperatures; will not bond to damp concrete surfaces; good abrasion resistance; poor freeze/thaw properties; toxic
Chlorinated rubber coating	Single or multi	Recoat after 10/15 years	No information available	Yes	Not applicable	Yes	No	Cannot use at low temperatures; pigmented

Table 3.2 *Summary of properties of surface treatments suitable for reinstating cover protection*

3.5.4 Cut out and reinstate

Where localised cover nonconformities have been detected, partial removal of the concrete element may be appropriate. The properties of the repair material chosen for reinstating the concrete to the required cover depth must be compatible with those of the original concrete and with structural or fire protection requirements[21,22]. Cutting out with tools can damage reinforcement, so consideration should be given to careful use of water jets. It should be borne in mind that this option can be aesthetically unacceptable since it is difficult to match the colour and texture of the reinstated material to the original.

3.5.5 Demolish and rebuild

Although the decision to demolish should not be taken lightly, in many cases this will be the only wholly acceptable solution. It may also be the most economical option. Low cover depths subject to severe exposure conditions are unlikely to be fully remedied by surface treatments. There is no entirely satisfactory substitute for an appropriate thickness of well-compacted, good quality concrete cover. Where long-term durability is of paramount importance and increased ongoing maintenance requirements cannot be tolerated, or where aesthetic considerations preclude the use of surface treatments, demolition and reconstruction may be the only recourse.

3.5.6 Financial settlement

There have been cases where demolition and reconstruction have been deemed to be warranted but impracticable for reasons of safety, convenience or programme. It is possible in some cases to agree on the likely effects of reduced cover in reducing the time before onset of corrosion and thus bringing forward the need for large scale maintenance. Based on reasonable expectations of the consequent future costs, it has been possible to enter into extra-contractual agreements allowing the nonconforming concrete to remain in place subject to an appropriate financial consideration.

3.6 PREVENTING RECURRENCE

3 6.1 Use of sufficient and appropriate spacers

One of the most common causes of nonconformities in cover is the use of insufficient or inadequate spacers. Research has shown that using at least two spacers per linear metre of reinforcement grid can dramatically reduce the risk of lack of cover[23,24]. In most cases where nonconformities in cover arise, the remedy is likely to lie in the use of more and/or better quality spacers. Comprehensive guidance is available on the use of spacers[24,25]. It should be noted that BS 8110[10] prohibits the use of site made spacers and that Concrete Society Report 101[24] recommends that concrete spacers be made of a minimum of grade C50 concrete.

3.6.2 Site supervision

Careful checking of the position and stability of reinforcement cages, together with the line and level of shuttering, before concreting is the best way to minimise occurrence of nonconformities in cover. It is good practice also to check dimensions of bars delivered to site against bending schedules. Spot checking after concreting, using a covermeter is also helpful as a policing quality control procedure, but the advantages of identifying and remedying problems before rather than after the concrete has been cast will be obvious.

3.6.3 Review site operations

Nonconformities in cover also arise as a result of site operations or systems, eg:

- inadequate formwork design

- as a result of disturbance to reinforcement by subsequent trades (including by concretors)
- disturbance to reinforcement during slipforming operations
- inadequate fixing of reinforcement or of formwork.

3.6.4 Review design

Certain aspects of the design of reinforced concrete structures are known to give rise to cover problems. These include:

- beam-column or beam-beam junctions where clashes of reinforcement lead to bars being fixed out of position
- rebates, holes, chases, penetrations, pockets, drip grooves, etc which are overlooked as areas where cover will be reduced
- at movement or construction joints, particularly where water bars are provided
- in slabs where reinforcement supporting chairs may be prone to collapse or twisting
- in the vicinity of cast in services (eg electrical conduits) where there is insufficient co-ordination between the various members of the design team
- where no or insufficient account is taken of the effects of tolerances in the bending of reinforcement and the construction of formwork
- congestion at laps.

Problems of this type can usually be avoided by paying closer attention to detailing and to the effects of the design on buildability.

3.7 TYPICAL EXAMPLES OF COVER NONCONFORMITIES ARISING FROM CASE REVIEWS

Example 1
The specified cover to the soffit of a large voided bridge deck was 35 mm. Covers as low as 8 mm were measured. It was believed that the placing of too stiff a concrete around the void had caused the problem. Various proposals for remedial action were discussed, and an offer was made to place a sum of money in a special bank account as a contingency against future maintenance problems. Finally, however, these were rejected and the bridge was demolished.

Example 2
A routine covermeter survey of the soffit of a glued segmental bridge indicated insufficient cover. It was believed that the spacers had broken under the weight of the reinforcement. A surface treatment was applied to enhance the protection to the reinforcement. Subsequent sections of the bridge were constructed using stronger spacers, but a covermeter check again indicated insufficient cover, similar to that of the original segments. The cover was then rechecked by drilling holes to the reinforcement. It was found that the correct cover had actually been achieved throughout and that the covermeter was faulty.

Example 3
Reinforcement for a culvert involved the use of dead fit bars (ie fitting between two formed faces without incorporating a lap). Insufficient allowance was made for tolerances in the construction of the formwork and the bending of the reinforcement. On placing the reinforcement in the shutter, the contractor realised that the cage was too large and the specified cover could not be achieved. The engineer amended the reinforcement details.

Example 4
A cover of 20 mm was specified to the top reinforcement in a 135 mm-thick concrete canopy. The steelfixer took the view that this would not be sufficient and so provided a cover of at least 40 mm. The canopy deflected excessively and had to be demolished.

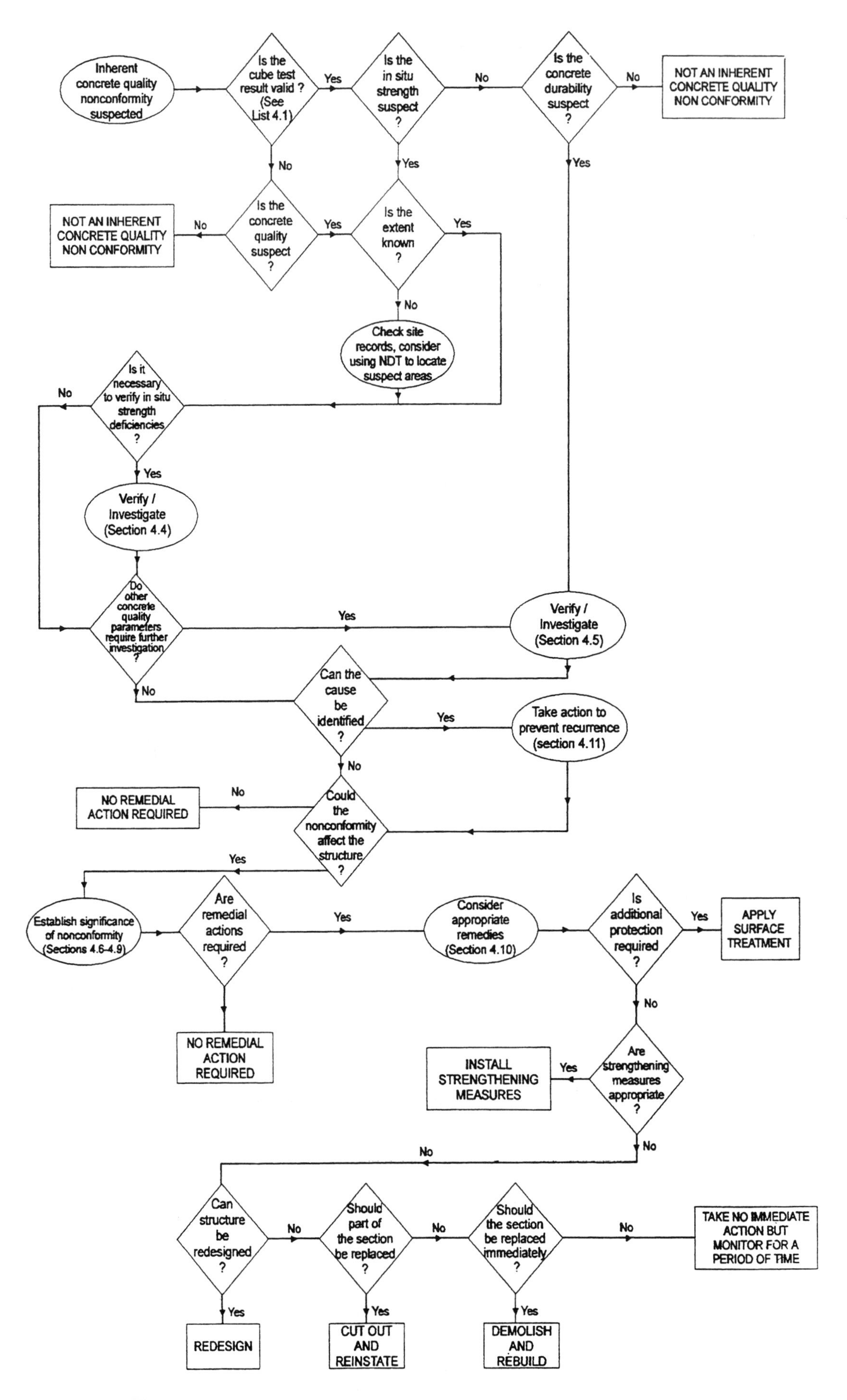

Figure 4.1 *Flow chart: inherent concrete quality*

4 Inherent concrete quality

4.1 INTRODUCTION

The decision-making framework for dealing with a suspected inherent concrete quality nonconformity is set out in the flow chart in Figure 4.1.

The essential approach to dealing with an inherent concrete quality nonconformity is once again to follow the four key processes set out in Section 2.

1. Verification/investigation.
2. Establishing the significance.
3. Determination of appropriate remedy.
4. Preventing recurrence.

When an inherent concrete quality nonconformity is suspected, the primary issue is likely to be strength, but durability parameters are also likely to require consideration.

4.2 VERIFICATION/INVESTIGATION: PRIMARY CONSIDERATIONS

The first indication of inherent concrete quality nonconformities is usually the failure of the standard concrete control cubes (as sampled, manufactured, cured and tested in accordance with BS 1881[26-29]) to meet the specified compressive strength requirements, although excessive retardation of the setting of fresh concrete may also indicate inherent concrete quality. When an inherent concrete quality nonconformity is suspected, there are three primary considerations underlying the verification/investigation process.

1. Establishing that the cube test results are valid.
2. Establishing the extent to which the cubes are representative of the concrete in the structure.
3. Determining the effects of shortcomings in strength on other concrete quality parameters.

The delivery tickets should be checked to ensure that the correct concrete was supplied and that the producer was not instructed to add additional water on site.

4.3 VERIFICATION/INVESTIGATION: ESTABLISHING THE VALIDITY OF CUBE TEST RESULTS

The concrete strength values obtained from routine 28-day compressive cube tests can be erroneous for a variety of reasons[26-29], including:

- sampling errors
- faulty cube moulds
- inadequate compaction or over-compaction
- cubes stored at temperatures outside the permitted range
- lack of moist curing
- operator errors during load tests
- faulty load test machines.

These errors can result in cube strength results that typically may be from 30 per cent lower to 10 per cent higher[30] than the strength values obtained for cubes that have been produced strictly by following the procedures set out in BS 1881[26-29].

When an inherent quality nonconformity is suspected, the sampling, cube-making and testing processes should be checked. A list of prompts for consideration when verifying the validity of cube test results is presented in List 4.1.

If possible, reference should also be made to the concrete producer's in-house cube test results, as these can provide a useful comparison. Once a cube test result has been verified as being below specification, all concrete associated with that batch and other associated batches are at risk. Site records should be used to establish the location of the suspect concrete in the structure.

Prompts for consideration when verifying the validity of cube test results

1. Was the cube making, storing and sampling carried out by a competent person, following the procedures set down by the relevant parts of BS 1881[26-28]?
2. Was the fresh concrete sampled properly?
3. Was the correct size mould used?
4. Was the mould damaged in any way?
5. Was the cube mould assembled properly and to the correct tolerances?
6. Was too much or too little mould oil used?
7. What was the temperature at the time of the cube-making?
8. What was the cube storage temperature for the first 24 hours?
9. What was the cube storage temperature thereafter?
10. Was the cube cured properly before demoulding?
11. What was the curing regime thereafter?
12. Was the cube continuously moist-cured?
13. Was the cube tested at the correct age?
14. Had the cube suffered any damage prior to testing?
15. Was the load testing carried out by a UKAS laboratory or other accredited organisation?
16. Was the cube compression strength determination carried out by a competent person following the procedures set down in BS 1881: Part 116[29]?
17. Was the cube properly located in the load test machine?
18. What was the load rate used in the test?
19. What was the mode of failure of the cube?
20. Is the laboratory testing equipment regularly calibrated?
21. Does the laboratory cross-check results with other laboratories?
22. Is the test statistically reliable?
23. Does the cube result meet the compliance clauses in BS 5328: Part 4[43] or other specification clauses?

List 4.1 *Prompts for consideration when verifying the validity of cube test results*

4.4 VERIFICATION/INVESTIGATION: STRENGTH DEFICIENCIES

4.4.1 Determination of degree and extent

When the location of the nonconforming batch of concrete is uncertain, there is some merit in using non-destructive testing techniques to investigate the variability of the in-situ concrete quality before embarking on more destructive techniques to determine actual strengths. Using these techniques, suspect elements can be compared with similar elements which are known to be sound.

Techniques suitable for providing an indication of the variability of the quality of hardened concrete are:

- rebound hammer (Section 4.4.2)
- ultrasonic pulse velocity(Section 4.4.3).

Techniques suitable for assessing in-situ concrete strength are:

- core testing (Section 4.4.4)
- in-situ load testing (Section 4.4.5).

Techniques suitable for assessing in-situ concrete strength, once they have been properly calibrated, are:

- penetration resistance (Windsor probe) (Section 4.4.6)
- pull-off (CAPO) (Section 4.4.7)
- other techniques (Section 4.4.8).

These techniques are briefly discussed below.

4.4.2 Rebound hammer (Schmidt hammer)

Undertaking a rebound hammer survey is the most common means of investigating the variability of the quality of hardened concrete.

Description of technique

Rebound hammers operate by impacting a known mass at a known velocity onto the concrete surface. On impact the mass rebounds, and the amount of rebound is measured and recorded as the rebound number. The extent to which the mass rebounds is an indication of the relative hardness of the concrete surface[31].

Spring-controlled plunger type rebound hammers are suitable for concrete strengths in the range of 20–60 N/mm^2. Pendulum-type rebound hammers are more appropriate for low-strength concretes.

Application and limitations

The rebound hammer provides a rapid means of investigating the variability of large areas of concrete in a short time. It does not directly measure in-situ concrete strength. It is a comparative technique which is well established and particularly suited to assessing concrete that is less than three months old. The rebound hammer should not be used to test normal concretes that are less than three days old. The technique is non-destructive, but small disc-shaped indentations can be left on the concrete surface, particularly if the outer surface is not fully hardened.

Rebound hammers are simple to operate, easy to use, light, highly portable and relatively inexpensive. Hammers can be operated horizontally or vertically. When using the rebound hammer technique to compare a suspect concrete with a sound concrete it is important to use the same hammer for each area. Rebound hammer settings can be adjusted for measurements on lightweight concretes and mass concretes. It is important that the hammers are regularly calibrated. The technique only measures the quality of concrete within 30 mm of the surface. It provides no information on the concrete beyond this outer layer. The technique does not measure in-situ concrete strength; it gives only an indirect indication of strength.

List 4.2 lists prompts for consideration when carrying out a rebound hammer survey.

Carrying out tests on a pre-determined grid is recommended, to reduce operator bias. Where large numbers of test results are obtained, the identification of areas of suspect concrete can be achieved more easily if the rebound hammer survey results are presented as contour maps. Alternatively, plotting the rebound numbers as histograms can verify poor construction methods or changes in concrete quality[4]. Interpretations of histogram shapes are shown in Figure 4.2.

Prompts for consideration when carrying out a rebound hammer survey

1. Is the concrete more than seven days old?
2. Is the concrete less than three months old?
3. Is the hammer setting appropriate for the concrete being tested?
4. Is the most suitable type of hammer being used?
5. Are calibration checks using a steel anvil done during the test programme?
6. Is the same hammer being used throughout the test programme?
7. If not, has the magnitude of the difference between the hammers been established for the specific concrete under test?
8. Is the concrete surface smooth and dry?
9. Are there any areas of honeycombing, open texture or surface damage visible which should be avoided during rebound hammer testing?
10. Does the concrete mix contain hard aggregates?
11. Is each recorded mean rebound number the average of at least ten readings?
12. Have all the readings recorded at an individual test point been included in the calculation of the mean rebound number?
13. Is a record being kept of which readings are in a horizontal and which are in a vertical direction?
14. Have edge areas been avoided?
15. Have sharp discontinuities been avoided?
16. Has any vibration or movement been detected in the concrete element during testing?
17. Are contour maps to be drawn?
18. Are histograms to be drawn?

List 4.2 *Prompts for consideration when carrying out a rebound hammer survey*

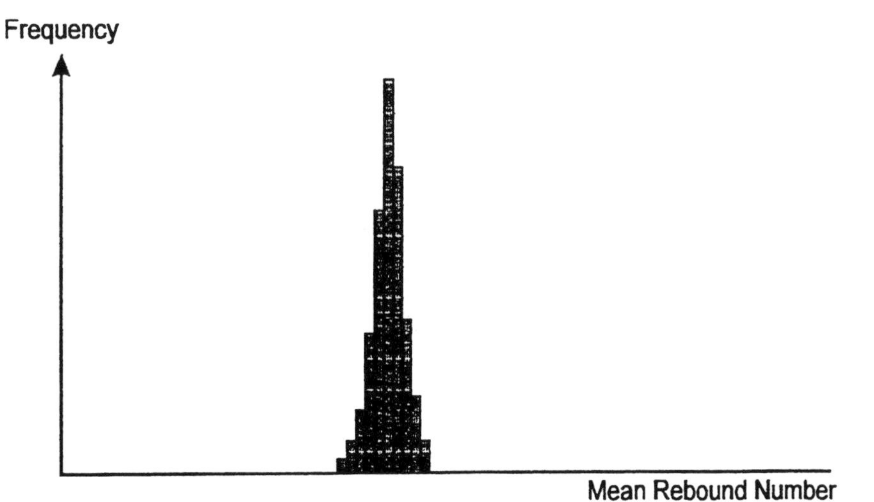

Single peak, narrow band suggesting well compacted, uniform concrete

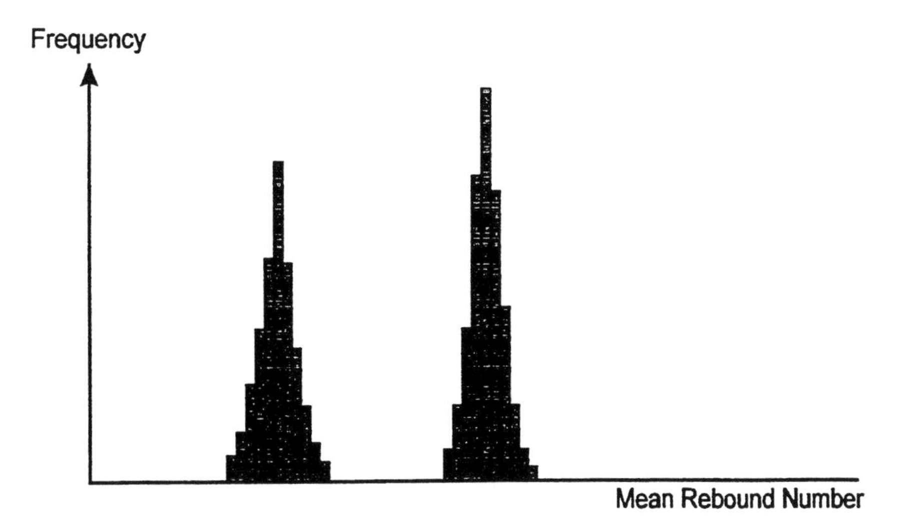

Twin peaks, narrow bands suggesting two different concretes, each of which are well compacted and uniform

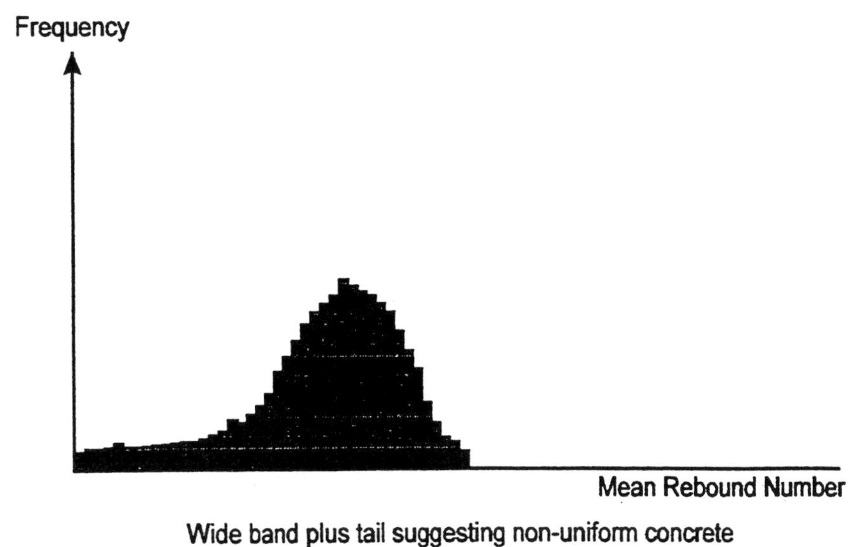

Wide band plus tail suggesting non-uniform concrete

Figure 4.2 *Histogram types obtained from rebound test programmes*

Reliability

The mean rebound number calculated for a test location should be based on a minimum of ten readings. In these circumstances, the accuracy of the technique can be taken to be $\pm 15/\sqrt{n}$ per cent, where n is the actual number of samples[31].

The accuracy of the results can be affected by:

- the presence of aggregates or steel close to the concrete surface
- the presence of any blowholes on the concrete surface
- any movement or vibration of the concrete element under test
- the slenderness of the concrete element
- the orientation of the hammer.

Attempts are frequently made to predict in-situ concrete strength from rebound hammer data, and manufacturers produce general strength correlations for use with their particular brand of hammer. These have limited practical value, however, and should be treated as indicative only. Although calibration trials for specific concretes in specific conditions can establish a relationship between rebound number and in-situ strength, these calibration trials are based on numerous test specimens covering a wide range of variables. It is rarely feasible in investigating nonconformities to produce sufficient test samples to validate a correlation between in-situ strength and rebound number. Ad hoc correlations between in-situ concrete rebound numbers and a small number of cubes or core specimens from the same concrete are unreliable. Rebound hammers should therefore be used only to monitor concrete variability and as an indicator of the best locations for other types of test.

4.4.3 Ultrasonic pulse velocity

The ultrasonic pulse velocity (USPV) technique is suitable for measuring in-situ concrete integrity by detecting variations in a concrete element. The technique can also detect internal voids and cracks (discussed further in Sections 5 and 7).

Description of technique

The USPV technique measures the time taken for a high-frequency pulse to travel through the concrete. If the path length between the transmitting and receiving transducers is known, the pulse velocity can be determined, which gives an indication of concrete quality.

Applications and limitations

USPV is a well-established technique for measuring the integrity of concrete. Like the rebound hammer, however, it does not directly measure in-situ concrete strength.

The range of pulse velocities measured on most concretes is narrow. Operating frequencies should normally be in the range of 50–60 kHz[6]. The path length should be at least 150 mm and must be established with precision. For best results the transducers should be set up on opposite faces of the concrete element, so that pulse travels directly through the element. Where access or size make this difficult, semi-direct measurement techniques can be used[9].

USPV equipment is robust, portable, easy to operate and widely available. Although USPV surveys need not be undertaken by specialists, they should be undertaken by well-trained operators and the results analysed by experienced interpreters. The equipment should be regularly calibrated and time-delay adjustments made[32].

It is better practice to obtain single readouts at several locations across a structural element and to produce a contour drawing to identify any suspect concrete, rather than to obtain several readouts at the same location. The technique should not be used on heavily reinforced sections because the pulse travels more quickly through the steel, and heavy reinforcement can therefore affect the pulse velocity. The technique is non-destructive, but localised staining of the concrete surface caused by the coupling medium may occur.

List 4.3 gives a list of prompts for consideration when carrying out USPV surveys.

Prompts for consideration when carrying out an ultrasonic pulse velocity testing survey
1. Is the testing being carried out by an experienced USPV operator?
2. Are the batteries fully charged at the start of testing?
3. Can the equipment measure transit time to an accuracy of ±1 per cent?
4. Can the path length be measured to an accuracy of ±1 per cent?
5. What is the pulse frequency?
6. Has the time delay adjustment been determined?
7. Has the time delay adjustment been checked regularly during the survey?
8. Is the concrete surface smooth and dry?
9. Has good acoustic coupling been achieved?
10. Is the separating layer thin?
11. Is the method of measurement direct, semi direct or indirect?
12. Is the path length greater than 150 mm?
13. Is the actual path length being measured or are nominal dimensions taken from the drawings being used?
14. If the path length is long, are signal amplifiers being used?
15. Has the equipment been calibrated against reference bars?
16. Are there any reinforcements bars in the pulse path?
17. Is the reinforcement heavily congested at the test points?
18. Are corrections being made to the pulse velocity values to account for the reinforcement bars?

List 4.3 *Prompts for consideration when carrying out an ultrasonic pulse velocity testing survey*

Reliability

Moisture conditions, mix characteristics, the presence of reinforcement, internal air-filled voids, internal cracks and poor acoustic coupling can all influence readings and reduce the reliability of the USPV technique. Errors in determining the path length can also have a significant effect on results. The reliability of the technique decreases for the semi-direct and indirect methods of measurement, because it is difficult to measure path lengths accurately in these situations.

Direct correlations between USPV readings and in-situ concrete strength are unreliable and should not normally be attempted. Calibration trials need to be based on numerous test specimens addressing a wide range of variables for a specific concrete. Such data may not be readily available when an inherent concrete quality nonconformity occurs, and the time and costs involved in setting up a retrospective calibration trial are likely to be significant. Even when such information is available, the accuracy of estimations of in-situ concrete strength based on ultrasonic pulse velocity measurements are unlikely to be better than ±20 per cent.

4.4.4 Core testing

Cutting cores from in-situ concrete is the most effective method for determining in-situ concrete compressive strength.

Description of technique

Cores, normally 100 mm or 150 mm in diameter, are cut from the hardened concrete and, after appropriate preparation, are tested by crushing in a laboratory. If required, samples from the crushed core can also be used for chemical analysis (discussed later). Estimations of in-situ cube strength can be made from core strengths using recognised formulae[33-35] (discussed further in Section 4.7.2).

Applications and limitations

Core testing is a well-established method for assessing in-situ concrete strength. Safe access for the core-cutting equipment must be provided. Core cutting is fairly disruptive, requires a supply of water and cannot be undertaken in low or freezing temperatures. The work should be undertaken by skilled operators.

The technique is destructive. A hole is made in the concrete, which has to be filled and can be unsightly.

Core locations have to be determined with care. Covermeters should be used to locate reinforcement bars before coring starts, and cores located away from reinforcement if possible. BS 6089[34] and Concrete Society Technical Report No 11[35] give guidance on the selection of core locations, but the need to obtain suitably representative samples will sometimes have to be balanced against the possible weakening of the structure by extracting cores from critical sections.

In general, a minimum of four cores per test location (eg per wall or per slab panel) should be taken. Core testing usually includes concrete density measurements and a visual examination of the internal concrete quality, as well as the strength determination[33-35].

Cores prepared for load testing should have a length-to-diameter ratio of 1.0 to 1.2.

Core testing is a moderately expensive technique.

Reliability

The accuracy achieved from core testing depends on the number of samples tested. A single estimation of in-situ strength can be taken to be accurate to within ±12 per cent. If n samples are taken from a given element, the accuracy improves to $\pm 12/\sqrt{n}$ %[4,34,35].

Consideration will usually need to be given to the relationship between the mean in-situ strength obtained from core testing and the characteristic strength required by the design.

4.4.5 In-situ load testing

Description of technique

In-situ load testing is carried out to assess the performance of a structure or element of structure under load. The behaviour of the structure, in terms of deflection and/or strain, and if necessary crack width, is normally monitored throughout the loading cycle[6,50].

Application and limitations

In-situ load testing is usually expensive and time consuming to carry out. Specialist advice may be required. It can be difficult to isolate the particular structural element to be tested. In-situ load testing should not normally be carried out on reinforced concrete that is less than 28 days old.

Appropriate safety precautions must be implemented to protect personnel and adjacent structures or parts of the same structure in the event of a collapse during the test. The member under test should, if possible, fail to safety, ie alternative load paths should be provided.

Load increments should be added and removed carefully. The maximum load may need to be left in place for a pre-determined period before being removed. Deflection and/or strain measurements should be recorded at each stage of the loading and unloading cycle. Load-deflection or load (stress) – strain plots are usually produced.

The choice of loading material should reflect the nature and magnitude of the required test load, the availability of materials and ease of access. Commonly used materials are water, bricks, sandbags and steel weights. Take care to ensure that the intended load cannot be inadvertently increased, eg by rainwater.

Extreme care should be taken in attempting to infer the behaviour of a structure under a higher loading than that to which it has been tested.

Reliability

Short term in-situ tests can give a reliable indication of structural adequacy provided that conditions are adequately controlled, including ensuring that the load is applied accurately; that an accurate datum is provided for the deflection measurements; that the member under test is free from unintended restrains, and that allowance is made for temperature effects where appropriate.

4.4.6 Penetration resistance ("Windsor probe")

Description of technique

A hardened steel alloy bolt (probe) is fired into the concrete using a powder cartridge. The length of bolt protruding from the concrete surface is measured to determine the length of bolt penetrating, giving an indirect indication of the concrete strength[6,36,37].

Applications and limitations

Penetration resistance tests can be carried out quickly for moderate cost. The technique is useful in situations where access for coring is difficult. Modern instruments are fitted with safety devices that prevent firing if the bolt is not perpendicular to the concrete surface.

This technique only measures the resistance of the surface zone of concrete from which in-situ strength estimations can be made. No information is gained about the bulk of the concrete.

Penetration resistance is destructive, leaving conically shaped areas of damage, typically 75 mm in diameter and 40 mm deep, at each bolt site. It is not suitable for testing high-strength concretes because insufficient bolt penetration is achieved. Larger bolts should be used for lightweight concretes. Slender concrete elements should not be tested using this technique, because of the risk of the concrete cracking and splitting.

Reliability

In-situ strength estimations should only be based on correlations prepared specifically for the concrete being tested. In these circumstances, the accuracy of the strength estimation can be taken to be ±20 per cent. Less reliable results are likely under site conditions. General strength correlation curves produced by manufacturers are empirical and often overestimate the in-situ strength, so should be treated with caution[36].

The test readings can be influenced by the presence of reinforcing bars and by cracking in the test zone. Low concrete strength and the presence of hard aggregate in the mix can also affect the accuracy of test readings.

4.4.7 Pull-out test (CAPO)

Description of technique

A compressed split ring is expanded into a groove undercut from a drilled hole in the concrete surface. The force required to pull out the ring, using a hand operated hydraulic jack, is measured. The maximum tensile load which can be applied to the ring is taken to be related to the tensile strength and also the compressive strength of the concrete[4,37].

Applications and limitations

The pull-out test is fairly expensive and time-consuming to carry out. Each test takes about half an hour, and a minimum of four tests are required per location. The testing should be undertaken by a specialist operator.

Installation of the split ring causes minor surface damage to the concrete.

The pull-out test only measures the properties of the surface zone of the concrete. No information is obtained with respect to the bulk of the concrete.

If the split ring is left embedded in the concrete after testing, localised rust staining and ultimately spalling of the concrete face can occur.

Reliability

Strength values determined from general correlation curves are empirically based and are unlikely to be more accurate than ±20 per cent. Under site conditions results are likely to be less reliable. General correlations can only be applied to normal-weight concretes. Where strength values can be determined from specific correlation curves produced for the particular concrete under test, the accuracy improves to ±10 per cent of the true strength value.

The reliability of this measurement technique is reduced by the presence of cracks or reinforcement bars in the test zone. Lack of care during the split-ring installation, or failure to apply the pull-out load at a steady rate, can also give rise to erroneous readings.

4.4.8 Other techniques

Other techniques for estimating in-situ concrete strength include:

- pull-off tests[32,37,38]
- internal fracture tests[6,37,39]
- break-off tests[6,37].

These are all surface tests which cause some surface damage.

The pull-off test is a relatively new technique, which requires careful preparation. In-situ strength predictions based on this technique can be accurate to ±15 per cent, but less reliable results are likely under site conditions.

The internal fracture test is a quick, relatively cheap technique with a high in-test variability. The likely accuracy of estimates of in-situ strength can be expected to be ±30 per cent, provided correlation curves specific to the concrete under test are used. Less reliable results are likely under site conditions.

The break-off test is a quick, moderately expensive technique, which also has a relatively high in-test variability. In-situ strength predications can be accurate to within ±20 per cent of the true strength value, provided that correlation curves are specific to the concrete under test. Less reliable results are likely under site conditions.

4.5 VERIFICATION/INVESTIGATION: OTHER CONCRETE QUALITY PARAMETERS

Before assessing the significance of a strength nonconformity, two further factors need to be investigated:

- whether and to what extent the nonconformity has arisen as a consequence of the mix proportions deviating from the specification,
- the likely effects of the nonconformity on the durability performance of the structure.

4.5.1 Mix proportions

Under current UK design codes[10,11,14] and standards for specifying concrete mixes[40-43], concrete mix proportions are specified in terms of minimum cement content, maximum water/cement ratio, and, if appropriate, cement composites (pfa, ggbs, etc). Inherent concrete quality nonconformities can arise as a consequence of batching errors or the inclusion of unspecified materials into the mix. Mix proportions can be verified by:

- observing the concrete batching process
- checking autographic records
- chemical analysis of the hardened concrete
- petrographic analysis of the hardened concrete.

These techniques are briefly discussed below.

Autographic records

Confirmation of the mix proportions may be obtained from the autographic records kept by the concrete producer. These should be checked if available, and will usually give a reasonably reliable indication of the mix constituents (but do not normally provide information on the water content of the mix).

Chemical analysis of hardened concrete

It is possible to undertake the chemical analysis of samples of the hardened concrete to estimate the initial cement content and water/cement ratio[44,45]. Experienced specialist analysts are needed to carry out this work.

Cement content analyses can be carried out on concretes containing Portland cement. Less reliable results are obtained for concretes containing ggbs and pfa.

Techniques for determining water/cement ratio are only suitable for normal Portland cement concretes. Water/cement ratios cannot be a determined for air-entrained concretes or concretes containing waterproofing admixtures. Determinations of water/cement ratios are difficult and expensive to carry out.

In favourable circumstances, the likely accuracy of a cement content determination is ±40 kg/m^3, while the accuracy of the original water/cement ratio determination is likely to be of the order of ±0.1. The reliability of the analysis results is affected by the aggregate properties, the degree of compaction, the presence of cracks, and any freeze-thaw damage. Single determinations of cement content are unlikely to achieve an accuracy better than ±100 kg/m^3.

The test results for chemical analysis of the concrete are generally highly dependent on the assumptions made by the analyst.

Petrographic analysis of hardened concrete

The identification of the concrete mix constituents or of any contaminants in the mix can be achieved by petrographic examination[45,46]. Specialist petrographers are needed and should be consulted before taking samples for analysis, in order to ensure that samples are properly taken and prepared.

Qualitative assessments can be made of aggregate type, cement type and any cement additions such as pfa and ggbs. Microsilica cannot be detected.

4.5.2 Durability performance

When an inherent concrete quality nonconformity is suspected, or has been confirmed, the potential long-term durability performance of the concrete may need to be investigated. Durability performance test methods are, however, still at research stage and can only be taken as indicators of future durability performance[47]. Specialist advice should be sought before embarking on investigations into future durability performance.

The long-term durability performance of concrete is known to be dependent on its ability to resist attack from aggressive agents such as water, carbon dioxide, chloride ions, sulphate ions and other chemicals. The rate at which these aggressive agents can ingress into the concrete is influenced by the permeation properties of the concrete.

In assessing the potential durability performance of a particular concrete, consideration should be given to the types of aggressive agent(s) and the transport mechanism(s) to which the concrete will be subjected, so that appropriate durability properties can be assessed. An indication of the durability properties of concrete can be inferred from measurements of:

- steady state water permeation
- non steady state water permeation
- capillary suction (water absorption)
- gas permeability
- gas diffusion
- ion diffusion.

Descriptions of laboratory test methods associated with these measurements of durability performance can be found in RILEM Report 12[48].

Details of in-situ tests which can be used to measure the relative permeation properties of the concrete surface can be found in Concrete Society Technical Report No 31[49].

4.6 ESTABLISHING THE SIGNIFICANCE OF CONCRETE QUALITY NONCONFORMITIES: INTRODUCTION

Nonconformities in inherent concrete quality must be assessed in relation to the significance of concrete quality *per se*, which is:

- the concrete must satisfy structural requirements, ie the structure must be capable of safely carrying the design load
- the concrete must satisfy durability requirements, such that the structure performs satisfactorily over its lifetime.

In the UK inherent concrete quality is normally specified[10,11,14,40-43] in terms of:

- minimum cement content
- maximum water/cement ratio
- minimum concrete grade.

In broad terms, the structural requirements are addressed by the minimum strength grade, while durability requirements are satisfied by limiting the minimum cement content and maximum water/cement ratio. Depending on what aspect of durability is being considered, concrete grade, water/cement ratio or the constituent materials may be the most important parameter. Frequently it is important to conform to all of these specified parameters. Where necessary, expert advice should be sought.

Inherent concrete quality may also be specified in terms of other parameters such as air entrainment, density, permeability or the quality of the constituent materials.

4.7 ESTABLISHING THE SIGNIFICANCE: STRENGTH

4.7.1 Concrete grade and characteristic strength

Reinforced concrete design is normally based upon a characteristic strength which is usually equivalent to the concrete grade specified. For example, if the characteristic strength assumed in the design is 35 N/mm^2, a grade C35 (or equivalent strength) mix would be specified.

Characteristic strength is defined as "the strength value below which 5 per cent of the population of all strength measurements are expected to fall". This definition is based on the assumption that the strength measurements follow a normal distribution, which (for other than very low and very high-strength concretes) is known usually to be the case. Implicit in any design, therefore, is the assumption that up to 5 per cent of the concrete in the structure may not be of the specified strength.

Design codes also recognise the fact that concrete cubes do not represent the true in-situ strength of the concrete. This is, first, because the way in which cubes are made and cured is not strictly representative of the way in which the concrete in the structure is placed and cured; and, second, because the way in which a cube behaves under test is not representative of the way in which the concrete in a structure behaves under load.

These differences are accounted for in part by the way in which the design equations contained in the codes are derived, and in part by the safety factors incorporated in the equations.

Thus, although the failure of cube test results to comply with the requirements of BS 5328[40-43] (or with other similar specification clauses) may constitute a non-conformity, the significance of the nonconformity must be assessed against the intentions and requirements of the design.

4.7.2 Specified strength and in-situ strength

Estimated in-situ cube strength, determined from core test results in accordance with BS 6089[34], is defined as "the strength of concrete at a location in a structural member which is estimated from indirect means and expressed in terms of specimens of cubic shape". A similar definition is contained in Concrete Society Technical Report No 11[35].

That is to say, the in-situ strength determined from cores should not be taken to be equivalent to or compared directly with the specified grade, since the concrete grade is defined in terms of a strength derived from idealised cubes tested in the laboratory, and not in terms of the actual strength of the concrete in the structure.

BS 6089[34], Concrete Society Report No 11[35] and the Institution of Structural Engineers *Appraisal of Existing Structures*[50] all take account of this in suggesting (slightly different but broadly consistent) acceptance levels for in-situ cube strength compared with specified strength. The differing rationales behind, and limitations on the use of, the various factors are explained in the relevant publications. It should be noted that other than in Concrete Society Report No 11[35], no specific guidance is given as to the number of in-situ strength test results to be taken into account in applying these acceptance factors. In practice, this will depend on the nature of the structural element being assessed and the perceived variability of the in-situ concrete.

4.7.3 Potential strength

Concrete Society Technical Report No 11[35] also contains recommendations for determining, from the core test results, the potential strength of the concrete. This is defined as "the notional strength of concrete considered as the average standard cube strength at 28 days for a single batch of concrete moulded wholly as standard cubes". The degree of uncertainty associated with the correction factors used to convert in-situ strength to potential strength is quite high, however, and estimates of potential strength are more likely to be of significance in addressing contractual arguments than in establishing the significance of nonconformities. Also, estimations of potential strength are only valid at present for Portland cement concretes, ie the formulae cannot be used for concrete containing cement composites. It is suggested, therefore, that for the purpose of establishing the significance of nonconformities, reliance should be placed on in-situ strengths only.

4.7.4 Design review

When a strength nonconformity is identified, the most common means of assessing its significance is by reviewing the original design.

This may be done either on the basis of the failed cube test results only, by downrating the characteristic strength used in the design in accordance with the actual cube results obtained on site, and then rerunning the design calculations to establish whether the shortfall in strength can be tolerated. Alternatively, if in-situ strength test results (most commonly from cores) are available, the design calculations can be reworked on the basis of the in-situ rather than the specified strength.

When reviewing a design based upon actual site conditions, considerable care is needed in the choice of partial safety factors to be used in the calculations. It will normally be over-conservative to apply the factors contained in design codes, because in a back-analysis or appraisal situation many of the uncertainties in respect of loadings and material strength that are present at design stage do not apply to the same degree once the structure exists. Detailed guidance is given in both BS 6089[34] and *Appraisal of Existing Structures*[50], and the rationale for the suggested values is explained in these publications. In particular, if the design review is based on in-situ strengths derived from cores, for the reasons set out in Section 4.7.2 above, it may well be appropriate to reduce the material partial safety factor from, say, 1.5 as given in BS 8110[10] to a lower value as summarised in Table 4.1.

Table 4.1 *Suggested values of materials partial safety factor (for concrete) to be used in appraisals based on in-situ strengths*

Guidance document	Suggested value for material partial safety factor for concrete
BS 6089[34]	not less than 1.20
Institution of Structural Engineers: *Appraisal of Existing Structures*[50]	
(i) for structures with well understood failure mechanism	1.25
(ii) for structures with less well understood failure mechanism or where members may fail abruptly	1.35
(iii) for slender columns	1.50

4.7.5 Corroborating data from different test methods

Wherever test results for in-situ concrete strength can be corroborated with data obtained from tests measuring other concrete properties, greater confidence can be placed on the in-situ strength data if similar patterns of inherent concrete quality emerge from the different test methods.

4.7.6 Cement composites

While cement composites concretes should be capable of readily achieving the required 28-day cube strength, concretes containing pfa and ggbs may gain strength more slowly in-situ than Portland cement concretes. Long term, however, these concretes achieve relatively higher in-situ strengths provided there is moisture available for continued hydration. It may, therefore, be appropriate to make allowances for this when evaluating the significance of low initial in-situ strength results. Further tests at later ages to confirm the achievement of long-term strength values may be appropriate as there is no standard method for quantifying the possible in-situ strength gains of these cement composite concretes. Heat of hydration effects may mean that in-situ concrete strengths are higher than the cube strength results.

4.8 ESTABLISHING THE SIGNIFICANCE: DURABILITY

4.8.1 Introduction

As discussed above, strength and durability properties are interrelated. In the event of cube failures, the primary concern is usually to establish whether the concrete strength actually achieved is adequate. The influence that any shortcomings in strength may have on other related aspects of concrete quality must, however, also be considered. That is, even if it can be shown that the nonconforming concrete has adequate strength for its intended purpose, the question of its ability to provide a sufficiently durable structure must also be addressed.

4.8.2 Exposure conditions

Inherent concrete quality requirements based on minimum strength grade, minimum cement content and maximum water/cement ratio are all linked to the severity of the exposure conditions[10]. Reference should be made to the exposure conditions assumed in the design of the structure. If it can be confirmed that the actual exposure conditions are less severe than assumed, a relaxation in the inherent concrete quality requirements may be acceptable, subject to the concrete continuing to satisfy the structural design requirements.

4.8.3 In-situ permeability

Unless permeability testing was part of the original contract it cannot be used to confirm an inherent quality nonconformity. However, permeability testing may provide a means of demonstrating the acceptability of the concrete. If the in-situ concrete can be shown to have a lower than expected in-situ permeability value, the concrete may be considered less vulnerable to the prevailing exposure conditions.

4.8.4 Cement composites

Lower in-situ permeability values and improved chloride and sulphate resistance properties are normally associated with cements containing cement composites[52]. These properties may be relevant in circumstances where lower in-situ strength values are acceptable for the structural requirements but the durability performance of the concrete is being questioned. Expert advice should be sought.

4.8.5 Trade-off

Before considering a trade-off between inherent concrete quality and exposure conditions, etc, as outlined in Sections 4.8.1–4.8.4, it is important to recognise that the specified bond and fire performance criteria must also be satisfied.

4.9 ESTABLISHING THE SIGNIFICANCE: OTHER CONSIDERATIONS

4.9.1 Consistency in quality of concrete

The consistency in the quality of supplied concrete can be monitored by calculating the standard deviation value associated with the available cube test results. Small values for the standard deviation suggest that the quality of the concrete supplied is consistent, although higher standard deviations, which are due to a step change in quality of the constituent materials, may still relate to consistent concrete supply.

Alternatively, by plotting cusum charts[53] or using simpler charts to log cube results[54], trends in the quality of the concrete supply can be monitored. On these charts downward slopes indicate declining concrete strengths, upward slopes indicate increasing concrete strength, and near horizontal lines indicate that consistent concrete strengths are being achieved.

Consistency in the quality of the concrete supplied can also be confirmed by non-destructive tests which show low variability in the quality of the in-situ concrete.

In establishing the significance of a lower than specified cube result, more confidence can be placed in the likelihood that this cube result is unusual and possibly the result of incorrect cube making and testing, if the overall concrete supply can be shown to be consistent.

4.9.2 Construction programme

The process of investigating nonconforming concrete quality can be lengthy, particularly if further investigations are called for. Once concrete quality is at risk, subsequent construction operations associated with that concrete may have to be suspended. Delays while awaiting test results are costly. Evaluations of the significance of a concrete quality nonconformity should address this. It may be more cost effective in the longer term to accept a small number of test results with a higher degree of uncertainty, and act upon these results, rather than pursue a more detailed and lengthy investigation.

4.9.3 Reliability of investigation technique

In establishing the significance of an inherent concrete quality nonconformity, reference should be made to the reliability of the investigation technique *per se* and to the level of uncertainty associated with testing[3].

4.10 DETERMINATION OF APPROPRIATE REMEDY

4.10.1 Introduction

Once an inherent concrete quality nonconformity has been established the remedial options are:

- take no action if the nonconformity is not significant
- take no immediate action but monitor
- provide additional wet curing
- delay applying critical loadings to the nonconforming element pending further strength gain
- provide additional protection through surface treatment
- implement strengthening measures
- redesign
- cut out and reinstate
- demolish and rebuild.

4.10.2 Take no action

If the shortcoming in the in-situ concrete quality is deemed not to compromise the structural adequacy of the structure or its longer-term durability performance, the nonconformity in quality might be tolerated and no further action taken.

4.10.3 Take no immediate action but monitor

In certain other circumstances it may be appropriate to take no immediate action. This situation may arise, for example, when the concrete shows slower than anticipated strength gain or when nonconforming concretes are considered to be adequate in terms of strength but have doubtful long-term durability qualities. Any decision to delay implementing a remedial action must, however, be based on a thorough understanding of the possible consequences.

Appropriate regular monitoring must be undertaken to confirm the hoped for strength gains or to detect any signs of early deterioration (carbonation, chloride ingress, etc).

Pursuing this option carries the risk that future remedial actions may be more costly, particularly if the concrete becomes contaminated in the interim or if a greater proportion of the structure is subsequently at risk.

The contractual implications of pursuing this option will need to be considered.

4.10.4 Additional wet curing

Additional wet curing or ponding[55] can sometimes be used to promote strength development in the concrete surface (ie cover) zone. This remedial action is only suitable, therefore, if the specification requirements are directed at the performance of the cover concrete.

4.10.5 Delay applying critical loading

In circumstances where the concrete has insufficient in-situ strength but there is confidence that the in-situ strength will increase sufficiently with age, it may be appropriate to delay applying critical loads to the understrength element until the requisite strength is achieved.

4.10.6 Providing additional protection

As discussed above, deficient concrete quality can increase the risk of reinforcement corrosion in the structure. This risk can be reduced by applying suitable surface treatments that prevent aggressive agents migrating through the concrete to initiate corrosion.

Surface treatments can be either penetrants or coatings. Penetrants form a barrier which is part of the concrete; coatings form a barrier on the surface of the concrete. Section 3.5.4 gives advice on selecting appropriate treatments and a list of prompts to assist in selection.

Applying surface treatments entails a commitment to regularly recoating the concrete for continued protection and for aesthetic purposes. It is important, therefore, to agree who will be responsible for these longer-term maintenance obligations and for their costs.

4.10.7 Strengthening measures

There are several strengthening measures which can be applied to enhance the load-carrying capacity of understrength structural concrete elements. These include:

- construction of additional supports
- use of sprayed concrete to enlarge the element
- plate bonding
- supplementary post tensioning.

Before adopting any of these measures – briefly considered separately below – it is important that checks be carried out to ensure that unacceptable stresses are not set up either in the nonconforming element or in the structure as a whole. Note that most strengthening measures will have some effect on the appearance of the structure.

Construction of additional supports

These can be ties, props, anchors or additional bracing, using steel or reinforced concrete sections. Entire structures with serious strength deficiencies have been successfully strengthened using combinations of these[56].

Use of sprayed concrete

Sprayed concrete is suited to strengthening slab soffits, beams, walls and structural linings. It can be applied using the wet or the dry process[57]. The wet process generally produces a more homogenous concrete with better frost- and salt-resistant properties[58]; the dry process can result in concrete with variable water contents but is easier to use in places where access is difficult. Only certified sprayed concrete operators should be employed. The finished appearance of the sprayed concrete may not be aesthetically acceptable. The application of a render may be required.

Plate bonding

Structural plate bonding is a means of providing supplementary reinforcement to concrete members by fixing thin plates to the concrete surface using resin based adhesives[59-62]. Plate bonding is suited to the strengthening of beams and slabs. It can also be used for floors and columns. The plates can be made from mild steel, stainless steel, carbon-fibre-reinforced plastics or aramid-fibre-reinforced plastics.

Achieving a good bond between plate and concrete is essential if an efficient structural connection is to be made. Trained personnel must be employed for this task and fire protection properties should be reinstated. Failure of the bond can occur; if so, the plate must be replaced. Appropriate access points for plate inspections and replacements should be provided. Choosing plate bonding as a remedial option carries with it longer-term responsibilities for periodically checking the bond between the plates and the concrete. It is important to agree who will be responsible for this longer-term maintenance care.

Supplementary post tensioning

Structural strengthening of reinforced concrete structures can be achieved by providing external prestressing, whereby prestressing tendons are installed around the concrete element and then tensioned[63,64].

4.10.8 Redesign

Where the in-situ strength of a concrete element is insufficient to support the required loading once the structure is completed, consideration may be given to redesigning the as-yet-unbuilt section to accommodate the lower in-situ strengths of the nonconforming concrete. For instance, if a beam is of insufficient strength, it might be possible to provide additional beams to relieve some of the loading on the deficient beam.

If this option is pursued, the costs of any design work, the possible delays to the construction programme, the additional costs of the modifications themselves and the altered appearance of the structure may need to be considered.

4.10.9 Cut out and reinstate

The partial or total removal of a concrete element may be appropriate if there is a localised concrete quality nonconformity. The repair material chosen for the reinstatement must be compatible with the properties of the original concrete and with any structural or fire protection requirements[65]. Cutting out with tools can damage reinforcement, so consideration should be given to the careful use of water jets.

This option might be aesthetically unacceptable since the repair material might not match the colour and texture of the surrounding concrete.

4.10.10 Demolish and rebuild

Demolition might be the only option when the inherent quality is so poor that the stability and reliability of the structure is at risk. Long-term durability considerations that cannot be satisfactorily remedied by surface treatments, or unacceptable ongoing maintenance costs or aesthetic considerations, can also make demolition and reconstruction the only option.

4.11 PREVENTING RECURRENCE

4.11.1 Cube making and testing

Suspected inherent concrete quality nonconformities frequently arise because of poor cube making and testing. Site staff should be trained to adhere strictly to the sampling, making, curing and testing requirements set down in BS 1881[26-29].

On-site weather conditions can affect the concrete sampling and testing processes[66]. Tolerances permitted for cube storage temperatures on UK sites[28] are 20±5°C. As ambient temperatures on UK sites are less than this in the winter and could be greater in the summer, it is essential to provide appropriate storage areas and tanks for cubes.

4.11.2 Review mix design

A review of the mix design should be undertaken when there have been cube failures or significant variations in results. Adjustments may be required to the cement content, to the water/cement ratio or to the type, size, grading and proportions of the coarse and fine aggregates. The proportions of any cement composite materials and the type and dosage level of any admixtures should also be reviewed. Any review of the mix design should also address the workability requirements of the fresh concrete to ensure that it can be handled and placed satisfactorily.

4.11.3 Review of batching process

Inherent concrete quality nonconformities can arise as a consequence of changes in constituent materials, incorrect batching of the mix constituents or incorrect doses of admixtures. These errors can occur at the concrete production plant or at the point of delivery. Review of the batching processes should address whether errors have arisen due to any breakdown in the automatic batching plant or as a consequence of human error or misunderstanding.

4.11.4 Supervision of concrete placing

Close supervision during concrete placing operations can minimise the occurrence of concrete quality nonconformities arising from unauthorised and uncontrolled additions of water. Concrete delivered in uncovered dumper or tipper trucks can receive extra water during heavy rain. This can be avoided by covering the fresh concrete with plastic sheeting.

The consistency of the fresh concrete can be checked by making visual inspections and testing at the point of delivery. Check that the producer is not being instructed by the purchaser to add water on site. Ensure that operatives place and compact the concrete so that it is fully compacted and there is no segregation of the mix. Remove any standing water before placing, or choose a placing technique that minimises intermingling.

4.11.5 Increase sampling rates

It is prudent to increase sampling rates at the start of the concreting work or when placing concrete in critical elements, because this will reduce the quantity of concrete at risk if the cube tests fail to meet the specified compliance standards.

4.11.6 Monitor concrete quality

Monitoring concrete quality by cube testing should be carried out during concreting operations. Monitoring concrete quality trends by using cusum charts[53], monthly average cube strength graphs[119] or charts logging cube results[54] allow changes in concrete quality to be identified quickly. These changes can then be queried and remedied before a serious concrete quality nonconformity occurs.

4.11.7 Specify additional cubes

Where the construction programme entails placing large volumes of concrete in a relatively short time period, specifying additional cubes for testing at earlier ages can provide an effective early warning system. It may be useful to carry out a trial test programme to establish the correlation between seven-day and 28-day strengths for the specific concrete mix, or to ask the concrete producer for this information. Results of seven-day cube tests will, of course, have no contractual status if the specification requirements are based on 28-day cube tests).

There is also merit in specifying additional cubes for testing at later ages, particularly for cement combination mixes. The later-age (56-day) cubes can then be used to confirm the anticipated longer-term strength gain (provided conditions are such that some further gain in strength of the in-situ concrete can be anticipated).

4.11.8 Analysis of fresh concrete

Where concrete construction operations proceed rapidly, such as in raft construction or slip forming, it may be appropriate to check the fresh concrete constituents before placing. Assessments of cement content and maximum water/cement ratio can be carried out using RAM (rapid analysis machine) or other appropriate procedures. Guidance on assessment methods is provided in DD83[130]. The analysis of fresh concrete should be undertaken by experienced personnel.

4.12 TYPICAL EXAMPLES OF INHERENT CONCRETE QUALITY NONCONFORMITIES ARISING FROM CASE REVIEWS

Example 1
A cube result of 26 N/mm^2 was obtained from concrete placed in an inverted T-beam highway deck, where a C40 concrete grade had been specified. After reviewing the compliance clauses the concrete was deemed not to conform. The contractor contacted the suppliers to enquire whether their cube results for the concrete were satisfactory, but the supplier was unable to track down the results. The designer was asked to check the design loading conditions and concluded that there was no spare capacity available. Core testing of the structure was proposed, and criteria for acceptability were agreed by the designer with other interested parties before coring commenced. Core tests showed that the in-situ concrete had estimated in-situ cube strengths in the range 50-70 N/mm^2. The engineer accepted the concrete in the works. Subsequent inquiries found that the cube-making had not followed the procedures in BS 1881 and, as the weather was hot, the cubes had been made and stored initially at temperatures of around 30°C.

Example 2
A concrete dam constructed some 800 m below ground in a coal mine was designed to withstand a head of water of up to 200 m in the event of an inrush. A low-heat, 30 N/mm^2 pfa concrete was specified. Some of the 28-day cube test results were found to have lower than specified strengths. Additional cubes for testing at 56 days had been specified, but these too failed to comply with the specified strength. Following discussions with the client and the contractor, core samples were extracted from the face of the dam. These were tested in the laboratory for compressive strength and for water permeability. Although C30 strength had been specified, this strength was not required to meet the structural requirements of the dam; concrete permeability was a more pertinent parameter. The core tests confirmed the lower in-situ strengths, but the permeability values were much lower than expected. Seepage calculations based on the in-situ permeability results indicated that the amount of water that could seep through the dam would be negligible. No further action was deemed to be required.

Example 3
A bridge abutment base had been constructed with a 70 per cent ggbs concrete mix. The characteristic strength was required to be 40 N/mm^2 at 56 days. The cube results at 56 days showed that only 30 N/mm^2 had been attained. By the time these cube results were available the abutment had been cast onto the base, so a large volume of concrete was at risk. The designer carried out a reappraisal of the structural design, which demonstrated that the lower in-situ strength was acceptable. However, the client was concerned, about the long-term durability performance of the concrete, so a concrete overlay and additional drainage were incorporated into the structure.

Example 4

An inherent concrete quality nonconformity was suspected in over 200 piles on a large piling contract for an out-of-town retail site. None of the 28-day cube results met the specified strength requirements. Investigations found that the cube-making and curing procedures had been incorrectly carried out throughout the project. Cores were drilled from the upper 2 m lengths of the piles, and the core specimens subjected to both compressive testing and chemical analysis to determine the cement content. Laboratory results showed that the in-situ strengths and the cement content were too low. Appraisal of the structural design of the piles established that the lower in-situ strength was adequate for the pile loading conditions, but concerns were raised about the long-term durability performance in the light of the lower-than-specified cement content. Consequently, permeation grouting was undertaken around the sides of the 200 piles.

Example 5

A bridge deck cantilever nib was required to achieve a characteristic strength of 50 N/mm^2 at 28 days, but the cube results indicated that the mean cube strength was 38 N/mm^2. The lower-strength value was satisfactory for the structural requirements of the bridge deck, but there was concern over the longer-term durability performance. Additional waterproofing was subsequently applied to the nib to address this concern.

Example 6

The floor slab in a large tunnel had been laid without taking any cube samples. The engineer required early strength data so as to decide whether to allow loading of the floor slab to proceed. It was agreed to test the near-surface strength properties of the floor slab using the CAPO (pull-out) test, since this would cause only minor damage to the slab and the CAPO test is known to be more reliable when used for early-age testing of concrete. The estimated in-situ strength values obtained form the CAPO test were deemed acceptable and the floor slab loading proceeded satisfactorily.

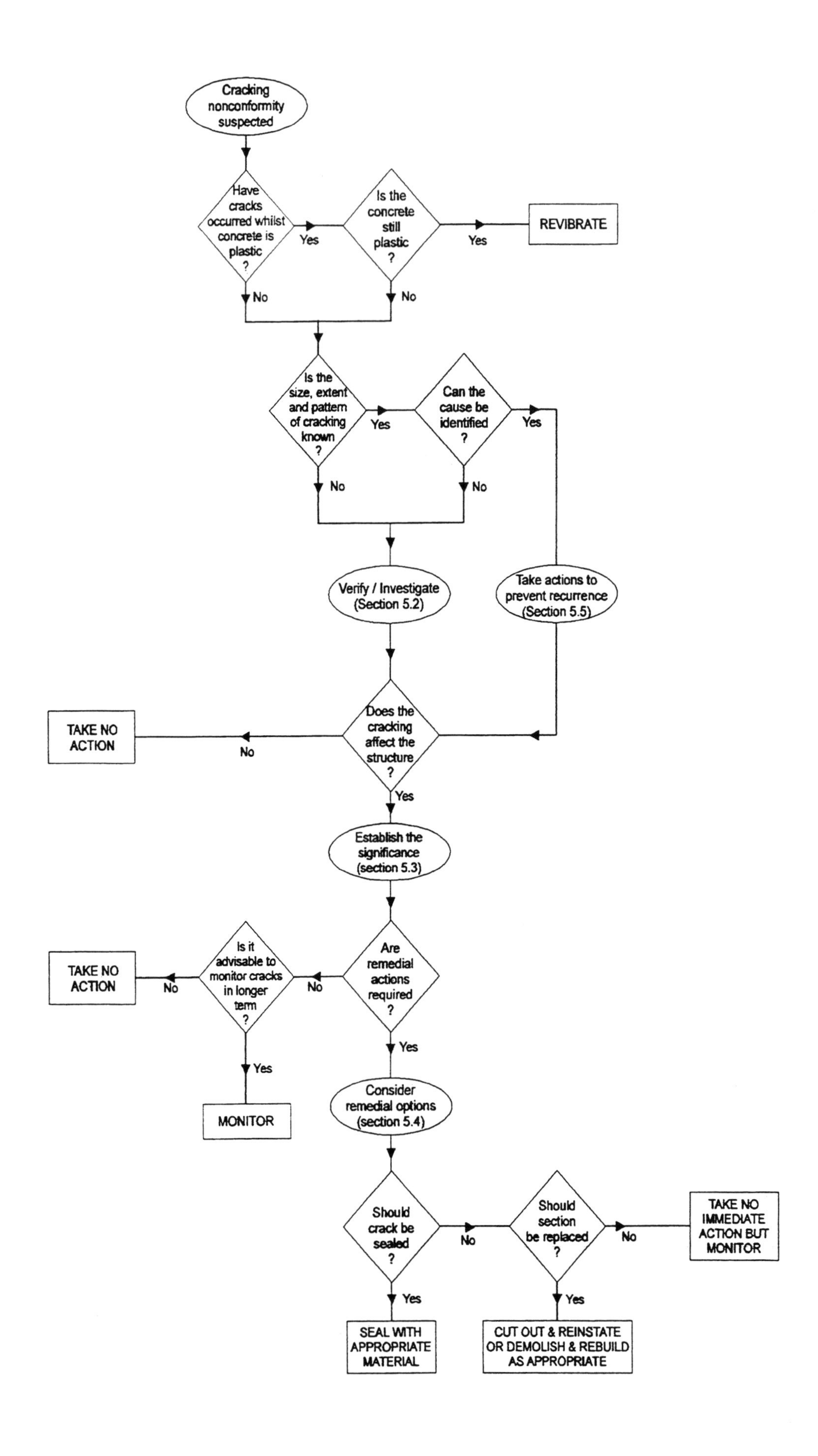

Figure 5.1 *Flow chart: cracking*

5 Cracking

5.1 INTRODUCTION

The section of the report is concerned with non-structural cracking in early-age concrete, and identifies the specific issues which need to be addressed when dealing with a cracking nonconformity.

The decision-making framework for dealing with a suspected cracking nonconformity is set out in the flow chart in Figure 5.1.

The essential approach to dealing with cracking nonconformity is once again to follow the four key processes set out in Section 2.

1. Verification/investigation.
2. Establishing the significance.
3. Determination of appropriate remedy.
4. Preventing recurrence.

The forms and causes of non-structural cracking are described in Concrete Society Report *Non-structural cracks in concrete*[67].

5.2 VERIFICATION/INVESTIGATION

Investigations into cracking nonconformities need to assess the location, spacing, orientation, extent, length, width, depth and pattern of the cracks. Any reliable information on the time at which the cracking appeared is also useful. Techniques available are listed below.

1. Detailed visual inspection (Section 5.2.1).
2. Crack width measurement (Section 5.2.2).
3. Crack mapping (Section 5.2.3).
4. Ultrasonic pulse velocity (Section 5.2.4).
5. Coring (Section 5.2.5).
6. Other specialist techniques (Section 5.2.6).

5.2.1 Detailed visual inspection

A detailed visual inspection can provide information on the location, spacing, orientation, extent and pattern of the cracking. Observable differences in crack width and crack depth should be noted but attempts to judge crack widths by eye should be avoided. The type of structural element in which the cracking has occurred should be noted (eg column, beam, wall, slab).

A detailed visual inspection is often carried out as a preliminary exercise. It should, nevertheless, be undertaken with care. A list of prompts for consideration when carrying out a detailed visual inspection of cracking nonconformities is given in List 5.1.

Prompts for consideration when carrying out a detailed visual inspection of cracking nonconformities
1. How soon after casting were the cracks noticed?
2. On which face(s) of the element are the cracks located?
3. What is the size of the element?
4. Is there one crack or several?
5. Do the cracks extend over the full height of the section?
6. Do the cracks extend through the full depth of the section?
7. What is the spacing between the cracks?
8. How long are the cracks?
9. Are the cracks narrow or wide?
10. Do the cracks seem to vary in width?
11. Are the cracks vertical, horizontal, diagonal, stepped or random?
12. Do the cracks form a pattern?
13. Can the cause of any restraint be identified?
14. Are the cracks located close to an edge?
15. Are the cracks located in or close to a change of section thickness?
16. Where are the movement joints in the section?
17. Do the cracks follow the line of reinforcement?
18. Is the cracking associated with the concrete surface which was in contact with the formwork?
19. Is the cracking associated with the position of the formwork tie bolts?
20. Is there any leakage, discolouration or efflorescence associated with the cracks?
21. What are the weather conditions (ie relative humidity, ambient temperature) at the time of the inspection?
22. What were the weather conditions subsequent to the concreting operations?
23. What are the lighting during the inspection?
24. What are the concrete mix design details for the cracked section?

List 5.1 *Prompts for consideration when carrying out a detailed visual inspection of cracking nonconformities*

5.2.2 Crack width measurement

Crack width measuring devices include:

- microscopes with graticules
- templates
- feeler gauges
- graduated rulers.

Description of technique

The crack width at the surface of the concrete is measured either optically or physically.

Application and limitations

Crack width measuring devices are inexpensive, portable, easy to use, and widely available. Crack widths can vary along the length of a crack. When this occurs, crack widths should be measured at several positions along the crack length. The best approach is to assess the overall widest part of the crack and then take a series of readings (say 3–5) along the length of the crack. The mean crack width and/or the crack width at the widest point should be recorded. Where widespread cracking is encountered it is not necessary or feasible to measure each crack. A representative sample of cracks should be selected. Crack width measuring devices only measure the crack width at the surface. Crack widths within the concrete cannot be measured using these devices.

The edges of cracks are often irregular, which makes it difficult to measure precisely the crack width. Flat feeler gauges should not be used to measure crack widths. It is better to use a wire of known diameter. Templates used in conjunction with a low-power microscope can increase the level of accuracy achieved.

Reliability

Crack widths can be measured to the nearest 0.05 mm. Measurements made using these width measuring devices are, however, to some extent subjective. No two observers should be expected to obtain precisely the same reading or repeatable results, particularly for fine crack widths. The reliability increases as the crack width increases. In view of this element of subjectivity, there is merit in categorising cracks in ranges of 0.05 mm from <0.05 mm to 0.5 mm.

Crack widths may also vary with the time, season and temperature at which they are measured. Consequently, this information should be recorded at the time the crack widths are measured.

5.2.3 Crack mapping

Crack mapping is a useful diagnostic tool. Crack mapping involves systematically noting the location, spacing, orientation, length and width of cracks. Cracks are mapped onto drawings or, if necessary, each crack can be marked with a dark coloured line and the overall crack map photographed. This latter method should be restricted to areas where the concrete is to be covered. Any pattern associated with the cracking can then be discerned.

Crack mapping is often supplemented with a covermeter survey to establish the location of the reinforcement. These details can then be mapped alongside the crack information.

5.2.4 Ultrasonic pulse velocity

The ultrasonic pulse velocity technique (USPV) can be used to estimate the depth of visible surface cracks.

Description of technique

The USPV technique measures the time taken for a high-frequency pulse to travel through concrete. In the vicinity of a crack the transit time is longer than in uncracked concrete since the pulse travels around the periphery of the crack. By measuring transit times for different transducer positions in the vicinity of the crack the depth can be determined[32].

Application and limitations

USPV equipment is robust, portable, easy to operate and widely available. Crack depth measurements, however, should be undertaken by experienced operators.

The distance between the transducers must be measured accurately. Crack depth determinations assume that there are no hidden cracks in the test zone. Water-filled cracks cannot be measured using digital readout USPV equipment. The USPV technique is better suited to measuring isolated well-defined surface cracks[68],and can be used to check whether the crack is perpendicular or oblique to the concrete surface.

Reliability

The accuracy of crack depth measurement is affected by the uniformity of the concrete in the vicinity of the crack. Moisture conditions in the concrete, the mix characteristics and poor acoustic coupling of the transducers to the concrete can lead to erroneous measurements of crack depth. Reinforcement bars can tend to short-circuit the signal and give erroneous readings for the crack depth measurement. Crack depths of less than 100 mm cannot be reliably established.

If the use of USPV is contemplated, there is some merit in calibrating or checking the measurements against cored samples.

5.2.5 Coring

The measurement of crack width and crack depth can be undertaken by injecting the crack with a coloured epoxy resin. Once the resin has set, the concrete can be cored and the width and depth of the resin measured. It should be noted that the use of resin is important. Crack width and depth measurements taken on cored specimens in which the cracks are unfilled can be unreliable since any stresses acting across the crack may be relieved on coring. The action of coring can also lead to cracks opening out.

5.2.6 Other specialist techniques

Other specialist techniques which can be used to detect cracks include:

- radar[9,70]
- acoustic emission[9,71].

These techniques are expensive and need to be undertaken by specialists.

5.2.7 Other considerations

The following information should also be obtained as they can assist in determining the causes of the cracking:

- the time at which cracking was first noted (but be wary of erroneous information)
- the temperature and humidity at the time of the investigation
- the temperature and humidity at the time of placing the concrete
- the curing regime
- try to locate the cause of any restraint as this is inevitably the basic reason for the cracking.

5.3 ESTABLISHING THE SIGNIFICANCE OF CRACKING

Concrete cannot be truly crack-free. Structural reinforced concrete is designed on the assumption that cracks may occur in the tensile zone when it is subjected to external loads. Other cracks arise as a consequence of natural movement against internal and external restraints.

The significance of cracking needs to be established in relation to:

- the type of cracking
- the influence of cracking on the durability of the structure
- the influence of cracking on the structural behaviour (eg deflections)
- aesthetic considerations.

Note that while a 0.2 mm structural crack may only open to this width when the structure carries the full design load; a 0.2 mm non-structural crack associated with early thermal movement or shrinkage is likely to be permanent; such a crack is potentially of greater consequence. Current design standards do not, however, address the significance of the difference between a temporary structural crack and a permanent non-structural crack.

5.3.1 Type of cracking

Distinctive crack patterns are associated with certain types of cracks. Crack patterns pertinent to early-age concrete are summarised in Table 5.1. Detailed descriptions of the various types of early-age cracking, their causes and the factors influencing their formation are well documented elsewhere[67,72]. Recognising crack patterns can assist in the identification of the causes of cracking.

Cracks in early-age concrete can also occur as a result of accidental damage, premature loading or the omission of reinforcing bars.

Combinations of different crack types can also occur, which can make establishing the causes of the cracking more difficult.

5.3.2 Exposure conditions

In circumstances where exposure conditions are particularly mild, eg dry internal environments, cracking might be tolerable from a durability viewpoint.

5.3.3 Crack orientation

Cracks that follow the direction of reinforcement are termed "coincident cracks". Those that cross the reinforcement are termed "intersecting cracks"[69,73]. Coincident cracks are more serious, being more likely to result in significant corrosion of steel reinforcement, since they may allow chlorides, carbon dioxide, moisture and oxygen to penetrate through the concrete cover to relatively large areas of steel. The corrosion reaction rates may also be affected. Isolated intersecting cracks are less likely to cause steel corrosion, but the presence of multiple intersecting cracks can increase the likelihood of corrosion[73].

Current codes and standards do not, however, distinguish between coincident and intersecting cracks.

5.3.4 Crack width and durability

Permissible values for crack widths in design codes are based on formulae which calculate a value which has a low probability of being exceeded[10,11,14,74]. Consequently, isolated instances of crack widths exceeding the permitted value by a small margin should not give rise to concern.

The considerations pertaining to permissible limits are related to the durability of the structure. Current permissible limits for crack widths used for design purposes in the UK range from 0.1–0.3 mm, depending upon exposure conditions[10,11,14]. These are summarised in Table 5.2.

Table 5.1 *Summary of crack patterns which occur in early-age concrete*[67, 72]

Type of crack	Pattern of cracking
Plastic settlement	Cracks become visible up to ten hours after placing. Usually noticed after one day. Cracks occur over horizontal bars; over formwork bolts; columns or narrow walls; at changes of depth in section. Cracks may have rounded edges. Cracks tend to be wide if not controlled by reinforcement.
Plastic shrinkage	Cracks become visible from one to eight hours after placing but are usually noticed after one day. Cracks are usually associated with slabs but can occur in the exposed top faces of walls. Cracks can be wide at the concrete surface (up to 3 mm) but rapidly decrease with depth. Cracks can form as diagonal cracks, 0.2–2.0 m apart; cracks can form a large random map pattern; cracks can follow pattern of reinforcement or changes of section. Cracks can pass through full depth of slabs but rarely reach the free edge of the slab.
Early thermal cracking	Cracks occur within two weeks. Cracks are associated with deep sections, thick slabs, sections where concrete is cast onto a previously hardened base (eg cantilever wall) or between sections without providing a movement joint. Cracks in walls generally begin at the base and extend about one-third of way up the wall. Cracks are likely to extend to the full depth of the section. Internal cracks may occur in deep sections.
Crazing	Cracks become visible within three weeks. Cracks are fine (0.05 mm) and form irregular map pattern areas up to 75 mm across on the concrete surface. The irregular areas can resemble hexagons. The cracks rarely exceed a few millimetres in depth. Cracks are usually more visible in concrete slabs that have been floated, or on formed surfaces, particularly for higher-cement-content mixes.
Drying shrinkage	Cracks become visible within one to six months. Cracks are usually associated with thin slabs and walls.

Limits on crack widths are based on the assumption that wider cracks are more likely to result in reinforcement corrosion than narrower cracks, for the same exposure conditions. However, for intersecting cracks there is little evidence to suggest that there is any relationship between crack width and the rate of corrosion of steel. For coincident cracks, corrosion of steel occurs irrespective of crack width[73,75].

It is important to distinguish between measurements of crack width for descriptive reasons and measurements of crack widths for specification compliance purposes. Note in particular that the crack width limits in BS 8007[14] are design values for use in calculations. These values should not to be used when judging whether or not a crack conforms with BS 8007, since conformity is checked by testing for watertightness.

Table 5.2 *Summary of maximum permissible crack widths recommended by design codes of practice*

Design code	Exposure condition	Maximum permitted crack width for design purposes
BS 8110:Part 1[10]	Very severe environment	0.30 mm
	Other environments	0.30 mm
BS 8007[14]	Severe or very severe environment	0.20 mm
	Critical aesthetic appearance	0.10 mm
BS 5400[11]	Extreme environment	0.10 mm
	Very severe environment	0.15 mm
	Severe environment	0.25 mm
	Moderate environment	0.25 mm
Eurocode 2: Part 1[74]	Class 2: humid environment	0.30 mm
	Class 3: humid environment with frost and de-icing salts	0.30 mm
	Class 4: seawater environment	0.30 mm

5.3.5 Crack depth

Cracks in the concrete cover to reinforcement may allow the ingress of aggressive agents which initiate steel corrosion. Crack depths which approach or reach the reinforcement are, therefore, more likely to initiate corrosion.

The widths of early-age thermal cracks will reflect the restrained contraction. Crack widths are likely to be widest in the centre of the section and narrowest where the reinforcement restrains the movement of the concrete.

5.3.6 Future changes in crack size

Seasonal temperature changes and the longer-term shrinkage of concrete can alter the size of cracks. Seasonal and daily temperature variations can cause cracks to open and close. Imposing live loads on a structure can also affect crack widths.

Estimations of future crack movements can be made by considering strains in the concrete arising from temperature changes, shrinkage or imposed loads[67].

Crack movements can be affected by the presence of a reinforcement bar located perpendicular to the crack.

5.3.7 Watertightness

Crack widths up to 0.2 mm in water-retaining structures may self-seal once they are exposed to water. It is usual practice to carry out water retention tests on these structures and to require any external leakages to be stopped where they exceed the specified limits[76]. In these circumstances, crack widths of 0.2 mm and less that do not show signs of sealing are deemed unacceptable and remedial measures are required[77].

Water excluding structures, such as basements, are designed and constructed to prevent water penetration. Hydrostatic pressure from groundwater can however force water via fine cracks in the concrete. As small amounts of leakage may not be as tolerable in water excluding structures, cracks must be effectively sealed, or other waterproofing measures, such as tanking, introduced. Guidance on current best practice is provided in CIRIA Report 139 *Water resisting basements*[128].

5.3.8 Aesthetic considerations

The significance of cracking from an aesthetic viewpoint should be evaluated in terms of crack width, viewing distance and the prestige rating of the structure[67]. Repairs can make cracks more conspicuous. Crazing type cracking can become ingrained with dirt over time, causing what may initially have been hairline cracking to become more noticeable. Deposits of calcium carbonate associated with fine cracks in structures subjected to water (including rain) that have been allowed to self-seal (through autogenous healing) can also be aesthetically unacceptable.

5.3.9 Subsequent finishes

Non-moving, non-structural cracks in concrete sections that are to be finished by applying tiles, render or other cladding materials should not normally give rise to concern (although drying shrinkage can cause the tiles or render to crack). Covering cracks with a finishing material can also provide protection to the reinforcement. Fixings should not, however, be located in cracks that may move after their installation.

5.3.10 Radiation shielding

In certain situations, concrete is used to provide radiation shielding. Any cracks in the concrete may reduce the effectiveness of the shielding.

5.3.11 Reliability of investigation technique

In order to establish the significance of a cracking nonconformity account must be taken of the reliability of the investigation technique *per se* and of the level of uncertainty associated with testing[3].

5.4 DETERMINATION OF APPROPRIATE REMEDY

5.4.1 Options and timing

Possible remedial actions for non-structural early-age cracking are:

- take no action
- allow cracks to heal autogenously
- brush dry cement into cracks and moisten
- seal/inject cracks with rigid sealant
- seal cracks using vacuum impregnation techniques
- seal/inject cracks with a flexible sealant
- apply a surface bandage
- undertake long-term monitoring of crack widths
- over coat for aesthetic purposes
- relax the specification
- cut out and reinstate
- demolish and rebuild.

If consideration is given to deferring remedial actions to suit the construction programme, it should be borne in mind that dirt and other detritus can lodge in the cracks in the intervening period, which might affect subsequent sealing and the longer-term durability and the appearance of the structure.

5.4.2 Take no action

Cracking may possibly be tolerated and no further action taken:

- where crack widths are as the design assumed
- where exposure conditions are unlikely to lead to corrosion
- where it is not necessary to guarantee the watertightness of the structure
- where subsequent finishes are to be applied
- where aesthetic considerations do not of themselves warrant remedial action.

5.4.3 Autogenous healing

In moist conditions, water can percolate through cracks. The water dissolves calcium hydroxide from the cement matrix and, when this calcium hydroxide solution comes into contact with carbon dioxide, crystals of calcium carbonate form, which can effectively seal the crack.

Autogenous healing is more likely to occur in sections that are in compression. Cracks greater than 0.2 mm are not likely to be sealed by autogenous healing[79]. It must be emphasised that cracks smaller than 0.2 mm may also not seal. Cracks that might undergo further movement are unlikely to be effectively sealed even if calcium carbonate crystals form.

5.4.4 Brush dry cement into cracks

Dry cement can be brushed into fine cracks and then moistened to form a cement grout, which effectively seals the fine cracks.

5.4.5 Seal/inject with a rigid sealant

If the cracking is not likely to undergo further movement it can be filled with a rigid sealant. Fine cracks are usually filled using injection techniques. Wider cracks on vertical surfaces are also usually sealed by injection, but wide cracks on horizontal surfaces can be sealed by cutting a "V" on the upper surface of the crack and placing the sealant material into the crack[80].

In choosing a sealant, consideration should be given to:

- the adhesion properties of sealant
- the shrinkage properties of sealant
- the resistance of the sealant to moisture and other aggressive agents
- the strength of sealant
- abrasion resistance of the sealant
- providing support for arrises
- the aesthetics of the repair
- cost.

Table 5.3 summarises the properties of the generic types of commonly used rigid sealants[69]. Specialist firms or trained, experienced operators should carry out this type of crack repair work.

Table 5.3 *Summary of properties of rigid sealants*[69]

Type of sealant	Mode of application	Properties
Epoxy resin Polyurethane Acrylic	Injection or pour into wider, horizontal cracks	• Good chemical resistance • good adhesion to concrete • non-shrink • rigid • low viscosity • gains strength in seven days • can be applied to damp concrete • can be injected into cracks 0.1 mm wide or less • good penetration • can be used to fill cracks up to 10 mm wide • concrete must be three weeks old • resin displaces water and debris from crack during injection • thioxtropic resins available for filling vertical cracks.
Cementitious grout	Pour into cracks or brush in dry cement and moisten	• Chemical resistance properties similar to concrete • can shrink • can take 2/3 weeks to gain strength • can be applied to damp concrete • suitable for cracks greater than 1 mm in width.
Latex emulsion or latex modified cement	Brush or pour into cracks	• Low viscosity • low elastic modulus • flexible • waterproof • cheaper than epoxy resins • better adhesion properties than cementitious-only grout • similar shrinkage to cementitious only grout • poorer chemical resistance than epoxy resins.

5.4.6 Vacuum impregnation

This technique is suitable for filling random multiple cracks[80]. It is quicker and less expensive than filling large numbers of cracks individually.

The area to be filled is enclosed with a plastic cover and then a partial vacuum is created before allowing a low-viscosity resin to flow in and fill the cracks. The technique is limited to concrete elements where the plastic cover and vacuum can be effectively installed. Specialist firms should be used. Subsequent coating of the concrete surface may be necessary for aesthetic reasons.

5.4.7 Seal/inject cracks with a flexible sealant

Flexible sealants should be used for cracks that are expected to undergo further movement. A recess is cut along the crack to reduce the amount of strain to be imposed on the sealant. The recess is then sealed using either a mastic sealant or an elastomeric sealant. The sealant must be fully bonded to the sides of the recess but should not bond to the bottom of the recess[80]. Table 5.4 summarises the properties of mastic and elastomeric sealants[81].

Table 5.4 *Summary of properties of flexible sealants*[81]

Type of sealant	Depth to width ratio for recess	Properties
Mastic or bituminous	3:1 2:1 1:1	• Can only withstand a small amount of movement • plastic • softens at higher temperatures (summer) • requires regular maintenance • expected life of 5–10 years • cheap • may require heat for application • can be used for vertical surfaces • dirt and debris can become embedded in the mastic.
Elastomeric (acrylic type)	2:1 1:1	• Flexible so suitable for moderate movements • can shrink • not suitable for cracks greater than 1 mm wide • can be used for vertical surfaces or lightly trafficked areas • no heat is required for application • concrete should be 14 days old • should be applied to dry concrete • expected life of 15 years • adhesion failure can occur if applied to wet surfaces.
Elastomeric (polyurethane, polyester or polysulphide types)	1:1 1:2	• Highly flexible • suitable for large movements • does not soften with time • expected life of 20 years • does not shrink • good adhesion properties to concrete • polysulphides and polyurethanes can be slow curing and need to be protected from damage until fully cured • failure to fully cure elastomeric sealants can lead to surface degradation on exposure to ultraviolet light (sunlight).
Silicone	1:2	• Highly flexible • suitable for large movements • expected life of 20 years • high initial cost • some formulations require particularly careful surface preparation.

In choosing a flexible sealant it is important to consider:

- repair
- the amount of movement that is likely to occur
- the size of recess that can be cut
- the location of the crack
- the adhesion properties of the sealant
- the elasticity of the sealant
- the resistance of the sealant to moisture and aggressive agents
- abrasion resistance of the sealant
- life expectancy of sealant and need for resealing
- the aesthetics of the repair
- cost.

Sealing is often best left until immediately before the structure is filled with water or otherwise covered, to minimise future movement of the crack and sealant. Once again, specialist firms or trained operatives should undertake this work.

Further guidance on sealant selection and sealing techniques can be found in the CIRIA publication *Manual of Good Practice in Sealant Application*[129].

5.4.8 Apply surface bandage

Surface bandages are strips of flexible sheeting, usually about 100 mm wide, which are bonded over cracks that are expected to undergo further movement. The edges of the bandage are sealed onto the concrete on either side of the crack leaving the central section free to move[80]. Surface bandages are commonly used in the following circumstances:

- when the crack is not in one plane
- when the anticipated movement cannot be accommodated by a suitably sized recess
- when cutting a recess is unacceptable.

Surface bandages can be masked by over-coating. It should be borne in mind that surface bandages bonded over cracks in water-retaining or water-excluding structures can be pushed off by hydrostatic pressure.

5.4.9 Long-term monitoring

When the cause of cracking cannot be determined, or when the long-term behaviour due to imposed loads, temperature variations or long-term shrinkage is uncertain, a programme of regular crack width measurement may be instigated and changes in crack width over several months monitored.

Monitoring screws or discs are fixed either side of the crack and the changes in distance between the screws measured. Demountable gauges or calipers are suitable for this purpose[82].

Telltales can be used to provide a rough guide to changes in crack width.

The contractual implications of pursuing this option will need to be considered.

5.4.10 Over-coating

Sealing cracks can result in the concrete structure being aesthetically unacceptable. Applying several layers of a suitable coating can provide additional protection to the concrete as well as producing a more pleasing surface finish. (For information on coatings refer to Section 3.)

Choice of suitable coatings must take into account the type of sealant used to fill the cracks. Incompatibilities between the coating and the sealant can give rise to debonding of the coating.

Opting to apply a coating to the concrete usually entails a commitment to long-term maintenance since the coating will probably require periodic renewal. It is important to agree who will be responsible for these longer-term maintenance commitments and, if necessary, set aside funds to meet the future costs of the works.

5.4.11 Cut out and reinstate

Localised cracking might necessitate the partial removal of a concrete element. Consideration should be given to the compatibility of the repair material with the remaining concrete as well as to the aesthetics. Cutting out with tools might cause damage to the reinforcement, so careful water jetting might be appropriate.

Whenever reinstated concrete pours are cast they are usually restrained more than the original concrete element. Where necessary, give consideration to new joints and to altering the reinforcement configuration.

5.4.12 Demolish and rebuild

Long-term durability considerations, aesthetic considerations, or the economic costs of repair and maintenance of the concrete over the lifetime of the structure, might make demolishing and rebuilding the only acceptable solution.

In rebuilding the section, consideration should be given to restraint effects acting on the reinstated concrete.

5.5 PREVENTING RECURRENCE

5.5.1 Review design: reinforcement provision

Crack widths can be controlled by providing sufficient steel reinforcement. For a given amount of steel, crack widths can be reduced by using smaller-diameter reinforcing. Extra steel will lead to finer (albeit more frequent) cracking.

5.5.2 Review design: restraints

Cracks occur when concrete is restrained, either in its plastic state or as a consequence of expansion or contraction in its hardened state. Consideration should be given to reducing the restraint associated with the following aspects of design and detailing:

- the provision and spacing of movement joints
- congestion of reinforcement
- the location of formwork tie-bolts
- cover depth
- sudden changes in depth of section
- boxouts
- curtailment of reinforcement
- casting sequence
- mix specification.

5.5.3 Review mix constituents

Plastic cracking is associated with bleeding in fresh concrete[67,83]. Plastic settlement cracking occurs when there is a large amount of bleeding, while lower bleeding rates increase the risk of plastic shrinkage cracking occurring. Adjustments of the mix constituents and mix proportions can reduce the amount of bleeding. Air-entrainment is an effective means of reducing bleeding. Details of other mix adjustments that can be made are well documented[67,83]

Choosing aggregates with low thermal expansion coefficients, reducing cement contents and using cement composites can all help to minimise early-age thermal cracking[72,84].

However, the use of cement composites to minimise early-age thermal cracking is not straightforward, since the cement composites can reduce the benefits gained from creep relief and early tensile strain capacity.

5.5.4 Weather conditions

Hot and cold weather can affect the bleeding of concrete, giving rise to plastic cracking[83].

Increases in the rate of bleeding can cause plastic shrinkage cracking. Covering fresh concrete with properly secured plastic sheeting held down securely at the edges is an effective means of preventing the occurrence of cracking due to the effects of high ambient temperatures and wind. Accelerating the set, altering the cement type or the use of sprayed curing membranes can also reduce the incidence of plastic shrinkage cracking.

In cold weather, concrete can bleed for comparatively long periods, which promotes plastic settlement cracking. Mixes containing ggbs are known to be prone to this type of cracking in cold weather. The use of air-entrainment can help prevent this problem.

Ambient temperatures can also affect the incidence of early-age thermal cracking. In hot weather, measures should be taken to reduce the temperature of the mix constituents[84]. In cold weather the temperature differential between the concrete surface and the surroundings is large. The use of insulation may be considered, but it is important to ensure that the insulation remains in place until the concrete pour has cooled[84]. Insulation can, in some circumstances, increase the amount of cracking because the insulation acts as an external restraint. Sudden removal of insulation may result in thermal shock to the concrete, which can also cause cracking.

Crazing is affected by low relative humidity conditions. In such conditions the use of continuous efficient curing can reduce the occurrence of crazing.

5.5.5 Revibration of plastic concrete

Plastic settlement cracking can be eliminated by revibrating the concrete. Revibration must be properly carried out and must be carefully timed since the concrete needs to be plastic enough to refluidise under the action of vibration but not stiff enough to leave a void when the vibrator is removed. Bleeding can recur if revibration is undertaken too early. During mild weather conditions, concrete can usually be revibrated up to three hours after placing[83]. Trowelling is inappropriate since the crack remains below the concrete surface.

Ensure site operatives are available to carry out revibration of the plastic concrete at the correct time. This is particularly important for the last pour of the day.

5.5.6 Review concreting operations

Pouring new concrete against hardened concrete can create an external restraint that can cause early-age thermal cracking. Typical examples are:

- a cantilever wall cast onto previously hardened base
- a concrete slab or wall cast against or between previously cast bays
- a casting sequence that did not minimise the number of trapped bays
- vertical lifts.

Providing movement joints or minimising the time interval between pours can minimise the amount of external restraint and reduce the incidence of cracking.

5.6 TYPICAL EXAMPLES OF CRACKING NONCONFORMITIES ARISING FROM CASE REVIEWS

Example 1
An 800 m^3 motorway bridge deck pour was placed during the month of May using an opc/ggbs concrete. On inspecting the deck two weeks after casting several large plastic settlement cracks were observed. Investigation revealed that the assistant resident engineer had allowed the contractor to place a number of loads of concrete at a higher workability than that specified, in order to complete the pour more quickly. The higher workability was achieved by adding water to the mix. Remedial works were delayed until September of the same year, when the cracks were filled with a cementitious grout. The contractor agreed to deposit a bond in a bank account to cover the costs of any further remedial works necessary in the 30-year design life of the bridge deck Future pours were restricted to a maximum of 200 m^3.

Example 2
A concrete tunnel below a river was cast using grade C40 concrete. A cofferdam had been constructed so that concreting operations could be carried out in dry conditions. The tunnel walls were 5 m high and 2 m thick. The formwork was stripped after 24 hours, revealing large thermal cracks in the tunnel walls. Subsequent investigations showed that temperatures within the core of the walls had exceeded 90 °C and a temperature differential of 60 °C had occurred between the concrete and the surroundings. The cracks were sealed by injecting them with epoxy resin. As the tunnel would be subjected to a 10 m head of water when in use, the contractor was required to construct an outer tunnel 500 mm thick, around the cracked tunnel.

Example 3
On this occasion, severe cracking occurred at the corners of a door opening in a reinforced concrete wall. Investigations established that a design fault had resulted in insufficient reinforcement being provided at the corners of the opening. The contractor was instructed to demolish the wall and rebuild it with the correct reinforcement. The cost was borne by the designer. (It is accepted that no amount of reinforcement can eliminate cracking at the corners of doors or boxouts, but providing sufficient reinforcement makes the cracks of less consequence.)

Example 4
Several plastic settlement cracks occurred at one end of a pretensioned concrete beam. The client initially wanted to reject the beam, but was persuaded to take expert advice by the contractor. The cracks were not structurally significant and it was intended, in any case, to coat the beams with waterproofing. The client subsequently agreed to seal the cracks by injecting them with epoxy resin. For other similar beams, the contractor altered the vibration sequence to prevent recurrence of the cracking.

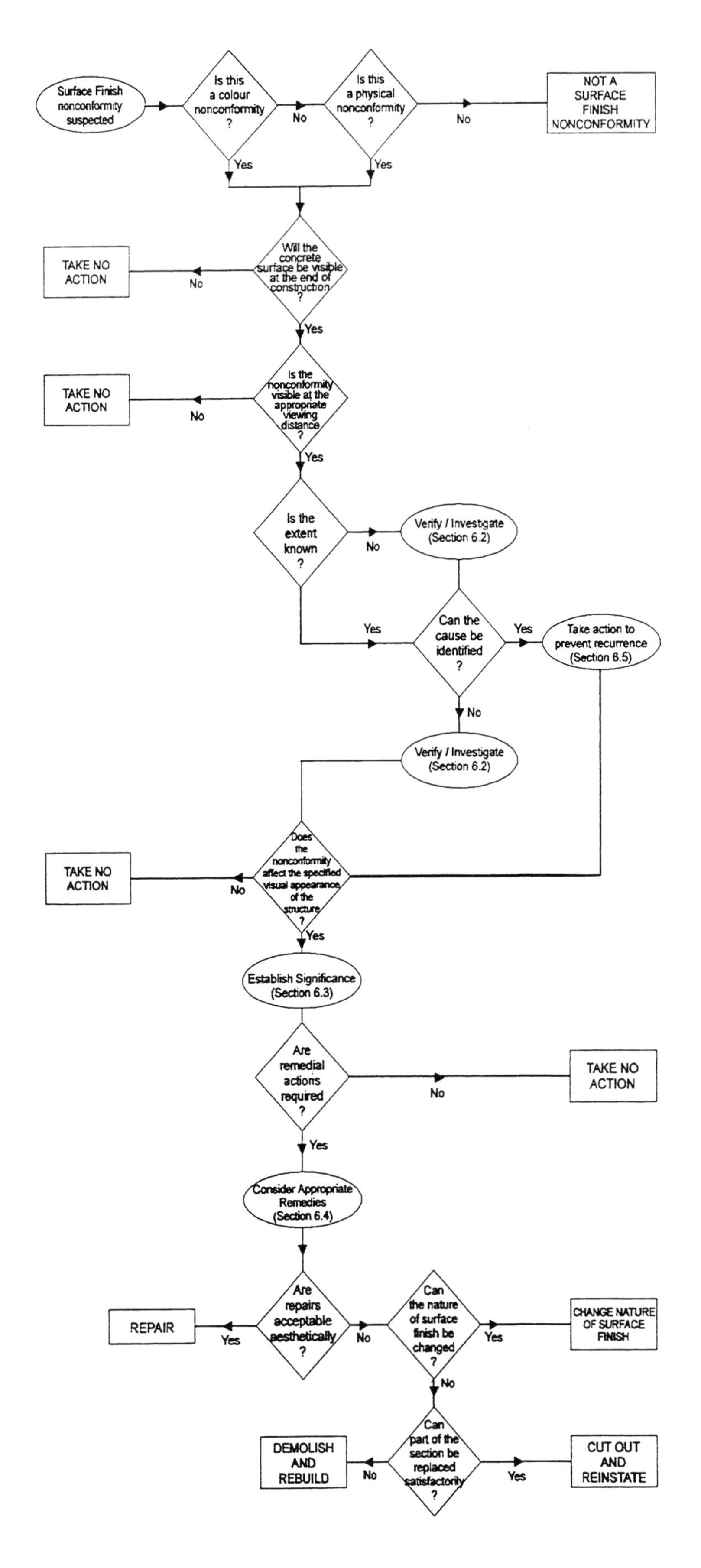

Figure 6.1 *Flow chart: surface finish*

6 Surface finishes

6.1 INTRODUCTION

The flow chart in Figure 6.1 sets out the decision-making framework for dealing with surface finish nonconformities.

The approach to dealing with a surface finish nonconformity is to follow the four key processes set out in Section 2.

1. Verification/investigation.
2. Establishing the significance.
3. Determination of appropriate remedy.
4. Preventing recurrence.

The major obstacle in assessing the quality of a surface finish, in normal practice, is the difficulty of adequately defining the required performance against which the suspected surface finish nonconformity can be judged. Assessments of the quality of a surface finish are rarely objective, and once a nonconformity has been confirmed it can be difficult to achieve an aesthetically acceptable remedy. This section of the report addresses the issues that should be considered when a surface finish nonconformity occurs or is suspected.

It is important to identify those who should view the surface finish when nonconformity is suspected. All relevant parties should be represented and they should view the surface finish nonconformity at the same time. Consideration should be given to involving a third party, such as an independent adviser (eg The Concrete Society Advisory Service).

Honeycombing and surface damage is covered in Section 7.

6.2 VERIFICATION/INVESTIGATION

6.2.1 Available techniques

Verifying that a surface finish nonconformity has arisen and investigating its extent is rarely straightforward. Assessment of the surface finish quality tends to be subjective. Where a surface finish nonconformity is suspected, there are two techniques available.

1. Detailed visual inspection (Section 6.2.2).
2. Quantification of the size and extent of the nonconformity (Section 6.2.3).

These techniques are discussed below.

6.2.2 Detailed visual inspection

Concrete surfaces are rarely blemish-free but the presence of significant blemishes gives rise to surface finish nonconformities. Comprehensive descriptions of 25 types of surface finish blemishes and their causes are given in *The control of blemishes in concrete*[85].

A detailed visual inspection can identify the type of surface finish nonconformity and its location on the structure. Surface finish nonconformities can be subdivided into two broad categories.

1. Colour nonconformities.
2. Physical nonconformities.

Table 6.1 summarises the blemishes associated with the above nonconformities.

Table 6.1 *Summary of surface finish nonconformity and associated blemishes*[85]

Surface finish nonconformity	Blemishes
Colour nonconformities	• Colour variation • Variations in tone • Mottled, flecked or speckled areas • Discolorations • Staining • Dusting • Efflorescence
Physical nonconformities	• Misalignment • Undulations on surface • Joint defects • Blowholes • Aggregate bridging • Grout loss • Grout leakage • Scouring • Scaling • Scabbing • Uneven aggregate exposure or distribution • Honeycombing (see Section 7) • Spalling (see Section 7) • Cracking (see Section5) • Loss of feature • Disruptions to pattern

While visual inspections may need to be carried out at close quarters to verify the type and extent of the nonconformity, consideration should be given to the extent to which the nonconformity would be evident from an appropriate typical viewing distance, so that only those nonconformities that are visible at the viewing distance normally associated with the structure are actually considered. Suggested typical minimum view distances[86] are given in Table 6.2.

A list of prompts for consideration when carrying out detailed visual inspections is presented in List 6.1.

Mapping surface finish nonconformities onto drawings or sketches or using marked-up transparency paper over photomontages is a convenient means of presenting the information. Photographs, properly taken and including a reference scale, can provide a useful, objective record of surface defects.

Table 6.2 *Viewing distances*[86]

Type of surface	Minimum viewing distance
Internal surface	1 m
External surface seen at close quarters	1 m
External surfaces of a building	3 m
External surface of civil engineering structures (including retaining walls and abutments)	6 m
Motorway bridge parapets; tall buildings or structures (not overlooked)	10 m (or more)

Prompts for consideration when carrying out a visual inspection to verify a surface finish nonconformity

1. Is the view blocked by falsework or scaffolding?
2. Will any areas be obscured during a later phase of the construction programme?
3. Will the section be clad at a later phase of the construction programme?
4. Is the surface finish smooth, textured, profiled, tooled, exposed aggregate or a combination of these?
5. Is the surface horizontal, vertical or sloped?
6. What is the appropriate viewing distance?
7. Does the concrete surface have a matt or shiny appearance?
8. Is the colour uniform?
9. Are the colour variations localised or in layers?
10. Are there sharp contrasts in colour variation or is the variation more gradual?
11. Are there any mottled, flecked or speckled areas on the concrete surface?
12. Are there any stains on the concrete surface?
13. Are these stains caused by aggregate contamination, associated with reinforcement or tie wire, or associated with release oil?
14. Does the concrete have a light-coloured, powdery surface?
15. Is there any efflorescence on the concrete surface?
16. Are there any blowholes in the concrete surface?
17. Are the blowholes isolated, in groups or spread across the whole section?
18. What is the maximum size and frequency of blowholes?
19. Are there any irregularly shaped cavities in the concrete surface?
20. Do any areas of the concrete surface have a coarse, stony appearance (ie honeycombing)?

List 6.1 *Prompts for consideration when carrying out a visual inspection to verify a surface finish nonconformity*

6.2.3 Quantifying the size and extent of a surface finish nonconformity

Suggested measurements to quantify the size and extent of the different types of surface finish nonconformities are listed in Table 6.3.

Table 6.3 *Measurements to quantify surface finish nonconformities*

Colour nonconformities	• Area • Range of colour variations observed
Physical nonconformities	• Depth/widths of misalignment • Depth of surface undulation • Joint width • Maximum size of blowholes • Number of blowholes • Area of blowhole group • Area of grout loss • Area of scabbing • Area of scouring • Area of spalling • Area or length of loss of feature • Length and width of damaged areas • Length and width of damaged ribs • Length and width of grout leaks • Distance between patterns

Measuring surface finish nonconformities is time consuming and can be costly, so it may be appropriate to restrict measurements only to those nonconformities that are visible at the appropriate viewing distance. It is important to relate measurements to a pre-agreed standard.

6.3 ESTABLISHING THE SIGNIFICANCE OF NONCONFORMITIES IN SURFACE FINISHES

6.3.1 Viewing distance

The significance of surface finish nonconformities should be judged at the appropriate viewing distance. Physical nonconformities become less noticeable as the viewing distance increases whereas colour nonconformities can become more noticeable.

6.3.2 Prestige of the structure

Prestigious buildings and structures, with fair-faced concrete surfaces, usually demand a high-quality surface finish. Nonconformities in surface finish are less likely to be tolerated in these structures.

6.3.3 Quality of surface finish

The quality required for surface finishes depends on the viewing distance and the prestige of the structure. The occurrence and extent of any surface finish nonconformity should be assessed within the context of the required surface finish.

Four classes of surface finish quality are proposed by CIB[87]. These are:

- special (highest standard of surface finish)
- elaborate
- ordinary
- rough (no requirements for surface finish).

For each class, limits are set for the magnitude of the surface finish nonconformity or for the amount of variation between different parts of the concrete surface[87].

This approach can be flexible. A different quality class can be selected for different types of surface finish nonconformity. For example, in a chosen section the colour variations may be required to satisfy the requirements of ordinary class whereas blowholes may be required to meet the requirements for elaborate class.

Establishing the significance of a surface finish nonconformity using this approach retrospectively may be difficult if the requirements were not defined in the specification.

6.3.4 Extent of surface finish nonconformity

Where the extent of the surface finish nonconformity is small in comparison to the surface area of the section, the nonconformity may be deemed acceptable. The extent of isolated surface finish nonconformities should be related to viewing distance. The tolerance limits proposed by CIB are for plane concrete surfaces[87], but this approach can be adopted for other surface finishes. Typical values for acceptable areas of isolated surface finish nonconformities for different classes of surface finish are given in Table 6.4.

Where isolated surface finish nonconformities are found to be regularly spaced, the resulting pattern may lead to the view that the nonconformity is acceptable.

Table 6.4 *Maximum permitted areas and equivalent diameters for isolated surface finish nonconformities (based on CIB recommendations[87])*

	Maximum permitted area of isolated surface finish nonconformity (mm^2) and equivalent diameter (mm)					
Type of surface	**Special quality finish**		**Elaborate quality finish**		**Ordinary quality finish**	
	Max area	Equiv diameter	Max area	Equiv diameter	Max area	Equiv diameter
Internal surface or external surface seen at close quarters	300	20	400	23	500	25
External surfaces of a building	900	34	1200	40	1500	44
External surface of civil engineering structure (including retaining walls and abutments)	1800	48	2400	55	3000	62
Motorway bridge parapets; tall buildings and structures (not overlooked)	3000	62	4000	72	5000	80

6.3.5 Unformed concrete surfaces

Unformed concrete surfaces have to conform to profile requirements and provide a smooth or lightly textured surface, depending on the class of finish stipulated [25]. In establishing whether a satisfactory unformed concrete surface has been achieved, consideration needs to be given to colour variations and to the presence of surface irregularities. In the absence of any specified criteria for assessing surface finish nonconformities in unformed concrete surfaces, it may be appropriate to adopt the tolerance limits set down for ordinary quality finishes by CIB[87], and to consider a minimum viewing distance of 6 m.

6.3.6 Loss of feature

Particular features of a surface finish, such as arrises or ribs, can be lost due to lack of care in tooling or formwork striking. Design features associated with profiled, textured or sculptured finishes may be too faint or may be omitted. In these circumstances, consideration should be given to the overall impression of the surface finish at the appropriate viewing distance. Minor loss of feature may be less conspicuous, or the loss of feature may be considered to be an acceptable variation to the design.

6.3.7 Disruption of designed pattern

Repeats of patterns at uneven rather than even spacings can constitute a surface finish nonconformity. When this occurs the overall effect should be judged at the appropriate viewing distance, rather than relying solely on measurements of pattern spacing. A single incorrectly spaced pattern can be unacceptable in what is otherwise a regularly repeated pattern, since it disrupts the symmetry of the facade; whereas several randomly spaced patterns can result in a less displeasing visual appearance that might be deemed acceptable.

6.3.8 Blowholes

Achieving blowhole-free concrete can be extremely difficult, and is not possible for vertical and sloped surfaces. Blowholes can occur in isolation, in groups or can be distributed over the whole section. In establishing the significance of blowholes, their size and their distribution need to be considered. Blowholes can be a problem when they occur in structures used for storing drinking water since the blowholes can harbour bacteria. Assessments of the incidence of blowholes can be made by reference to a series of photographs produced by CIB[87] and the degree of variation observed then related to the required quality of the surface finish.

Widespread coverage of the concrete in small blowholes is likely to be more acceptable than several isolated large blowholes, since the former will be less conspicuous when viewed from an appropriate distance. Small isolated blowholes can usually be ignored. Groups of blowholes can be considered as localised surface nonconformities and their significance judged in terms of area (see Table 6.4).

Small blowholes in textured and in exposed aggregate surface finishes are less conspicuous than similar-sized blowholes in a smooth surface finish. However, the incidence of large blowholes in textured, tooled or exposed aggregate surfaces are usually more noticeable and can result in a loss of feature. Blowholes may become more obvious with acid-etching or grit-blast finishes.

6.3.9 Weathering

External surfaces are altered by weathering. Differences in the amount of dirt accumulating on the concrete surface, rain washing down the concrete surface, and wetting and drying effects can all cause colour variations. Weathering effects are usually more noticeable on smooth finishes[88]. Detailing is important to prevent surface water running down the face of the structure. Within a few years, weathering can result in a wider range of colour variations on the concrete surface than may be apparent at the time of construction. This could influence any decision regarding action in respect of the nonconformity.

6.3.10 Cover requirements

Surface finish nonconformities such as blowholes can reduce the amount of effective cover to reinforcement in localised areas. This loss of cover needs to be assessed (see Section 3). Tooled finishes and exposed aggregate finishes can also lead to reduced cover depths if insufficient extra cover is not provided before finishing.

6.3.11 Obscuring the concrete surface

Wholly or partly obscuring the concrete surface with pipes, fire escapes or other structural elements later in the construction programme can help to mask surface finish nonconformities. In these circumstances the nonconformities may be less significant.

6.3.12 Applied finishes

If the concrete surface is to be clad or rendered, surface finish nonconformities are unlikely to be a problem unless they affect strength, durability or dimensional requirements.

6.3.13 Smooth concrete surfaces

Large areas of smooth, plain concrete inevitably contain colour variations[85,88]. Gradual changes in colour are usually preferable to sudden contrasts.

6.4 DETERMINATION OF APPROPRIATE REMEDY

When a surface finish nonconformity has been established the remedial options are:

- take no action
- apply a coating or other finish to mask the nonconformity
- make good any localised surface finish nonconformities
- change the nature of the concrete surface
- apply cladding
- relax specification requirements
- cut out and reinstate
- demolish and rebuild.

6.4.1 Take no action

When the surface finish nonconformities are minor, or where they are found to be less conspicuous at an appropriate viewing distance, and if the nonconformity does not compromise the required quality of the surface finish, it might be possible to tolerate the nonconformity and take no further action. Some colour variations lessen with time, so taking no further action could be a viable option.

6.4.2 Aesthetic considerations

Remedial works to surface finishes are seldom successful aesthetically, the repairs sometimes turning out to be more unsightly than the original nonconformity. One of the reasons for this is the difficulty associated with achieving a satisfactory colour match between the repair and the existing concrete, even when the same mix constituents are used. Other reasons include differences in curing and differences in surface texture between the hand-applied repair and the cast concrete. With time, these differences can become accentuated and the repair can become even more obvious and unsightly.

6.4.3 Coating

Colour variations and shallow physical surface finish nonconformities can be masked by applying one or more coats of paint, render or other decorative finish[85,89]. However, applying a finish can emphasise certain types of surface finish nonconformity (eg blowholes) and it can be difficult to achieve a good colour match between the finish and the surrounding concrete. It may, therefore, be necessary to coat the entire section. Coatings usually need to be regularly reapplied, which has obvious cost implications. It is important to agree who will be responsible for these longer-term maintenance commitments and, if necessary, to set aside funds to meet the future costs of the works.

Before applying a coating or other finish it is prudent to undertake trial samples to demonstrate the likely appearance of the treated surface.

6.4.4 Make a feature of the repair

Where repairs are necessary for durability purposes; such as filling blowholes, filling tie bolt holes, repairing damaged arrises, etc; or where remedial actions are necessary to improve the appearance of the concrete surface, such as making good grout leaks along construction joints; consideration may be given to making a feature of the repair. Adopting this approach can sometimes result in a more aesthetically acceptable solution than would be achieved by trying to hide the repair. (An example of this technique is the use of patch repairs of regular size, standing in relief of the concrete surface.)

6.4.5 Making good

High standards of workmanship, achieved by the proper use of materials applied by operatives experienced in surface finish repair techniques, are normally essential in making good surface finish nonconformities. Details of other making good techniques can be found in the BCA pamphlet *Concrete on site: making good and finishing*[89].

6.4.6 Changing the nature of the concrete surface

Many surface finish nonconformities are located in the outer 5 mm of the concrete. Removing this outer surface can thus effectively eliminate the nonconformity. Grit-blasting or tooled finishes can consequently be considered as remedial options. It is prudent, though, to prepare samples to demonstrate the resultant finish before embarking on this option.

Tooling or light grit-blasting can emphasise surface finish nonconformities like grout leakages but can effectively remove small blowholes and minor misalignments at construction joints, and deep point tooling removes most surface finish nonconformities. Tooling and light grit-blasting can also emphasise any cracking in the concrete, since these operations abrade the crack edges[72]. Heavy grit-blasting exposes the coarse aggregate and can, therefore, alter the colour of the concrete surface. Also removing the concrete surface can expose the spacers.

In addition, it is most important, before deciding to change the nature of the concrete surface by removing part of the original surface, that consideration be given to cover requirements. Grit-blasting and tooled finishes can remove up to 5 mm of the concrete; deep-tooled finishes can remove as much as 20 mm.

6.4.7 Cladding

Where the quality of the surface finish is unacceptable, consideration may be given to cladding the section, but dimensional requirements and aesthetic considerations should be addressed before pursuing this option. Guidance on the wide variety of available materials and techniques for cladding external walls can be found in the CIRIA publication *Wall technology*[127].

6.4.8 Cut out and reinstate

Opting to cut out defective areas and reinstate the concrete can remedy a localised area of deep surface finish nonconformity. Matching the colour and the surface texture of the repair with the existing concrete is difficult, however, and in the longer term the reinstated area can become increasingly noticeable as a result of differences in weathering.

6.4.9 Demolish and rebuild

This may be the only acceptable solution, particularly where high-quality finishes are demanded. Before choosing this option, however, bear in mind that if the specification requirements are particularly onerous similar problems may recur. The extent of demolition also needs to be carefully considered, in that partial reconstruction can be evident and can of itself have an adverse visual impact, possibly more so than the original nonconformity.

6.5 PREVENTING RECURRENCE

6.5.1 Review specifications

Specifications for surface finishes should describe as objectively as possible the quality of surface finish required. Specifications should avoid subjective phrases such as "uniform colour" and "blemish-free surface". The approach adopted in BS 8110[10] of type A, B or C surface finish has its limitations (see Section 6.5.11). References to the quality of surface finish on existing structures can be helpful, as can setting tolerance limits for surface blemishes along the principles set out in the CIB Report[87]. Where the quality of surface finish is important, trial panels should be made before construction. These can verify the construction method and can be used as a benchmark for the acceptability of the finishes achieved during construction (see Section 6.5.10).

6.5.2 Review design

Certain aspects of the design of reinforced concrete structures can contribute to surface finish nonconformities that can sometimes be mitigated by paying attention to detail and to buildability. Consideration should be given to:

- whether composite construction is feasible, using precast panels possibly as permanent formwork to provide the desired surface finish
- the provision and location of construction joints
- avoiding congested reinforcement that does not allow sufficient space to compact the fresh concrete properly
- the provision and location of secondary reinforcement
- the use of colour-matching spacers
- avoiding complicated details
- changing the type of surface finish.

6.5.3 Review mix design

To achieve a satisfactory surface finish may require adjustments to be made to the cement content, aggregate proportions and aggregate grading. The approach required in designing a concrete mix to achieve a quality surface finish differs from that required to achieve a given compressive strength grade. Basic guidelines are available[88]. Variations in colour may be less obvious if white cements are incorporated in the mix.

6.5.4 Supply of materials

A consistent supply of materials can reduce the incidence of colour variations in the surface finish. This is particularly important for the cement, any cement composites and the fine aggregates, since it is these materials that contribute most to the concrete colour. Colour variations can be caused by variations in the carbon contents of pfas. Contaminated aggregates (eg those containing pyrites) should be avoided.

6.5.5 Formwork

The quality of surface finish is highly dependent on the formwork[91]. To reduce the incidence of surface finish nonconformities, consideration should be given to:

- the most appropriate type of formwork material (timber, steel, grp, etc)
- the absorbency of the formwork lining
- the number of times formwork is reused
- the cleanliness of formwork
- preventing water collecting in the formwork
- preventing damage to forms during handling
- storing formwork correctly
- wearing down the surface of steel or grp forms before using them
- replacing formwork linings frequently to maintain the quality of the finish
- ensuring that release agents are applied thinly and uniformly over the whole formwork lining
- properly supporting formwork
- securing and properly sealing joints in formwork
- position of joints in formwork (as these are usually evident on the concrete surface)
- preventing formwork from drying out
- appropriate striking times
- not using formwork that has been treated with retarders, other than where exposed aggregate finishes are intended.

6.5.6 Controlled permeability formwork

The use of controlled permeability formwork may be appropriate in some circumstances, as this produces a concrete surface with a slightly textured, matt finish and few blowholes. The colour of the concrete is generally darker than that obtained with impermeable formwork, but the overall colour can be more uniform subject to particular considerations and selection of particular types of this formwork. Expertise is necessary and reference should be made to specialist suppliers, literature and the CIRIA Report[92] on this subject.

6.5.7 Site operations

Surface finishes are sensitive to small differences in mix ingredients or placing techniques. Close supervision of the following site operations, by competent and experienced staff, can effectively reduce the incidence of surface finish nonconformities:

- formwork construction
- batching
- mixing
- transporting
- placing
- compacting
- forming construction joints
- formwork striking
- curing.

6.5.8 Pour sizes

Although large volume pours can eliminate the need for construction joints, the surface finish obtained is more likely to show colour variations, often in layers. Overall improvements to the quality of the surface finish may be achieved by limiting the pour size, but thought must be given to the siting of construction joints, otherwise the joints between lifts can detract from the quality of the finish.

6.5.9 Protruding reinforcement

Rusty starter bars protruding from the top of a recently cast section can lead to staining of the concrete when rain washes down the face of the section. This can be prevented by covering the protruding reinforcement or by shaping the top of the concrete section to prevent rainwater running down the concrete surface. Advice on the protection of starter bars can be found in CIRIA Report 147[93].

6.5.10 Trial panels

Trial panels can be used to confirm construction techniques and to identify any problems associated with the design details or the concrete mix. Trial panels should be:

- produced under site conditions
- full size (if possible)
- constructed, as far as possible, in exactly the same way as the actual construction
- at the same orientation (vertical, horizontal or sloped) as the type of panel or surface which they are intended to represent.

Once a satisfactory trial panel has been produced it can serve as a reference for the acceptability of the actual construction.

6.5.11 CONSTRUCT panels

CONSTRUCT evaluation panels are being established at six centres around the country. These panels contain locally available materials produced to BS 8110[10] Type A and Type B surface finishes. It is intended that they be used as references for describing surface finishes. More details can be obtained from CONSTRUCT (tel 01344 725 744).

LEEDS COLLEGE OF BUILDING
LIBRARY

6.5.12 Access for finishers

Grit blasting, tooling and the production of exposed aggregate finishes should be carried out by experienced finishers. Providing adequate access ensures that the equipment is used correctly, without the operators needing to reach or stoop.

6.5.13 Visually prominent sections

Unless absolutely necessary, visually prominent sections should not be the first cast. Rather, their construction should be deferred until any teething problems have been addressed.

6.6 TYPICAL EXAMPLES OF SURFACE FINISH NONCONFORMITIES ARISING FROM CASE REVIEWS

Example 1
The ceiling of a prestigious building consisted of a beam grid. Lighting fitments were to be placed in the recesses between the beams. Collapsible formwork was used, the formwork face being a smooth resin-faced plywood. On striking the formwork the concrete surface was found to be discoloured, due to hydration effects. This is a typical problem associated with smooth, impermeable form faces. This nonconformity was satisfactorily resolved by applying a hand-stoned finish on the beams and altering the light fittings.

Example 2
Satisfactory trial panels were produced before start of construction of a large car park. However, on striking the grp formwork during construction the concrete surface had a blotchy appearance. The car park comprised large bays and the contractor had elected to use large pours with surface areas in excess of 500 m^2. This volume of concrete took several hours to place, and the large pours resulted in longer vibration times than had been used for the trial panels. The appearance of the concrete was deemed to be unacceptable. The car park was rebuilt.

Example 3
A ribbed finish was required for a bridge abutment. On striking the formwork numerous blowholes were found in the ribs. The blowholes were the result of inadequate vibration of the concrete in the ribs. Achieving adequate vibration was difficult since the ribs' cross-sections were too narrow to accommodate poker vibrators and the size of the shutters meant that external vibration was ineffective. The problem was resolved by the contractor. A small poker vibrator was inserted into a metal tube attached to the outside surface of each rib, providing sufficient vibration to reduce the number of blowholes to an acceptable level.

Example 4
A grit blast finish was specified for a bridge abutment. The grit blasting was satisfactorily completed but within a few days the concrete surface was covered in brown stains. Subsequent investigations revealed that the coarse and fine aggregates contained ironstone, which had been exposed by the grit blasting. The ironstone had then oxidised causing the staining. Remedial actions involved wire brushing the concrete surface to remove the worst of the staining and then coating the surface. The requirement to grit blast the concrete surface was deleted from the specification.

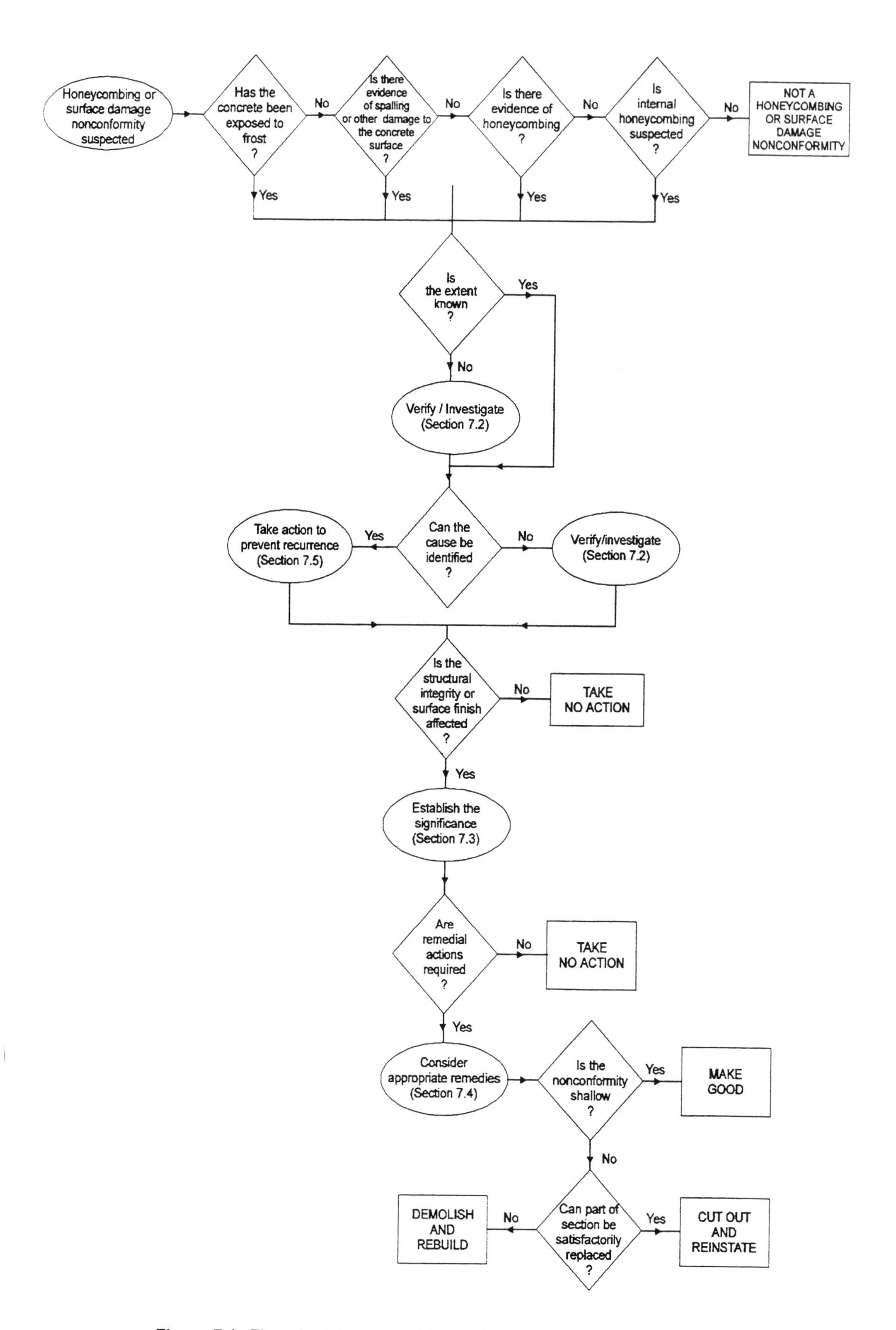

Figure 7.1 *Flow chart: honeycombing and surface damage*

7 Honeycombing and surface damage

7.1 INTRODUCTION

The decision-making framework for dealing with honeycombing and surface damage nonconformities is set out in the flow chart in Figure 7.1.

The basic approach adopted for dealing with honeycombing and surface damage nonconformities follows the four key processes set out in Section 2.

1. Verification/investigation.
2. Establishing the significance.
3. Determination of appropriate remedy.
4. Preventing recurrence.

The implications of any honeycombing and surface damage in a reinforced concrete structure need to be addressed in relation to aesthetics; loss of cover; adequacy of reinforcement protection; and the actual or effective loss of member cross-section.

7.2 VERIFICATION/INVESTIGATION

Honeycombing is the phenomenon whereby coarse, stony concrete is produced, either on the concrete surface or internally, as a consequence of poor compaction, mix segregation or lack of fines. Honeycombing can be visually unacceptable. This aesthetic aspect needs to be addressed separately from the effect of honeycombing on the structural integrity and durability of the concrete element. Surface damage refers to:

- spalling of concrete
- scaling of the concrete surface due to freeze-thaw action
- accidental damage caused by impact or abrasion.

Techniques that can be used to verify that a honeycombing or surface damage nonconformity have occurred and to investigate their extent are listed below.

1. Detailed visual inspection (Section 7.2.1).
2. Ultrasonic pulse velocity (Section 7.2.2).
3. Coring (Section 7.2.3).
4. Other specialist techniques (Section 7.2.4).

These are discussed below.

7.2.1 Detailed visual inspection

A detailed visual inspection should be undertaken carefully, noting the location and extent of any surface honeycombing or surface damage. Since these affect the appearance of the surface finish of the concrete, the visual inspection should give consideration to the appropriate viewing distance (see Section 6). There may also need to be supplemented by checks on the reinforcement cover depth in the vicinity of the honeycombing or surface damage (see Section 3). Checks on the variability of the quality of the concrete surrounding the areas of honeycombing or surface damage may also be required. It may be appropriate to undertake at least a preliminary check on this, using a rebound hammer, at the same time as the visual inspection (see Section 4).

A list of prompts for consideration when carrying out a visual inspection of honeycombed or surface damaged concrete is presented in List 7.1.

Prompts for consideration when carrying out a visual inspection of honeycombed or surface damaged concrete
1. Is the surface of the concrete coarse and stony (ie honeycombed)? 2. What are the size of the voids in the area of honeycombing? 3. Does the concrete surface show evidence of scaling? 4. Is any concrete missing or spalled? 5. Where is the area of honeycombing or surface damage located? 6. What is the areal extent of the honeycombing or surface damage? 7. Does the honeycombing or surface damage extend through the full depth of the section? 8. Does the concrete in the vicinity of the honeycombing or surface damage appear to be sound? (Check with a rebound hammer.) 9. Is there any honeycombing associated with surface damage? 10. Is there any honeycombing associated with horizontal joints? 11. When was the formwork stuck? 12. Did any damage occur while striking the formwork? 13. Has any damage occurred due to collision, a blow, abrasion or other accidental cause? 14. Has the concrete been exposed to freezing conditions? 15. Is any reinforcement visible in the area of honeycombing or surface damage? 16. Is the honeycombing or surface damage visible at the appropriate viewing distance?

List 7.1 *Prompts for consideration when carrying out a visual inspection of honeycombed or surface-damaged concrete*

7.2.2 Ultrasonic pulse velocity

The ultrasonic pulse velocity (USPV) technique can be used to estimate the depth of any surface honeycombing or scaling of the concrete. It can also be used to detect any areas of internal honeycombing.

Description of technique

The USPV technique measures the time taken for a high-frequency pulse to travel through concrete. Transit times are longer in areas of honeycombed or poorer-quality concrete[32].

Applications and limitations

The USPV equipment is robust, portable, easy to operate and widely available. Although localised staining of the concrete may occur, the USPV technique is nondestructive. The detection of areas of honeycombing or scaling should, however, be undertaken by personnel experienced in carrying out USPV surveys.

To detect the depth of surface honeycombing or scaling, transit times are measured as the transducers are moved incrementally closer or farther apart across the concrete surface (ie utilising indirect transmission)[32].

To detect any areas of internal honeycombing, the transducers should be placed on opposite faces of the concrete section (ie utilising direct or semi-direct transmission)[32]. Drawing a grid on the face of the section and taking measurements at the grid points can assist in locating the area of internal honeycombing. Measurements can be made at points criss-crossing the grid lines. The grid size should reflect the size of internal honeycombing which it is considered would adversely affect the structure.

The path length between the transducers must be accurately measured. The distance between the transducers and the type and size of transducer can affect the magnitude of the honeycombing or surface damage which can be detected.

Reinforcement affects the pulse velocity since the pulse travels more quickly in steel (see Section 4).

The detection of honeycombing and scaling nonconformities is based on comparing the pulse velocities obtained in the suspect concrete with the pulse velocities in unaffected areas. The technique assumes that the concrete surrounding the area of honeycombing or surface damage is uniform.

Reliability

Moisture conditions, mix constituents, the presence of reinforcement, the presence of cracks and poor acoustic coupling can reduce the reliability of the USPV technique. Errors in determining the path length also affect the results. Areas of surface honeycombing or surface damage that are less than 100 mm in depth cannot be reliably detected; nor can areas of internal honeycombing that are less than 100 mm in diameter. The reliability of the results depends on path length and grid spacing.

7.2.3 Coring

The depth of any surface honeycombing can be determined by drilling cores through the affected area and measuring the extent of the honeycombing in the cored specimen[35,50].

7.2.4 Other specialist techniques

Internal honeycombing can be detected using specialist techniques such as:

- radar[94,111]
- radiography[95].

These techniques may be useful in certain specific circumstances, but they are expensive and the investigations must be undertaken by specialists; the data obtained will also require expert interpretation.

7.3 ESTABLISHING THE SIGNIFICANCE OF HONEYCOMBING AND SURFACE DAMAGE NONCONFORMITIES

7.3.1 Viewing distance

The significance of any surface damage should be assessed at the appropriate viewing distance; any nonconformities that are found to be inconspicuous can usually be ignored, subject to cover requirements being satisfied.

7.3.2 Loss of cover

Surface damage can result in the loss of concrete whereas honeycombing results in a poorer-quality concrete. In both cases the amount of cover provided to the reinforcement is effectively reduced. The amount of cover loss and the areal extent of any reduction in cover needs to be evaluated. Small isolated instances of reduced cover arising from honeycombing or surface damage may be acceptable provided that the location and exposure of the concrete to aggressive agents is not critical (see Section 3).

7.3.3 Exposure conditions

Reference should be made to the exposure conditions assumed in the design of the structure. If the actual exposure conditions are less severe than assumed, the consequences of the honeycombing or surface damage may be considered acceptable. In making this decision, however, it should be remembered that honeycombed areas can increase the risk of frost attack and may allow the ingress of aggressive agents into the concrete. Honeycombing increases the amount of concrete surface that may be prone to chemical deterioration, such as sulphate attack.

7.3.4 Bond between concrete and reinforcement

Internal honeycombing can result in reduced bond between the concrete and the reinforcement. The structural design consequences of any reduction in bond strength may need to be evaluated.

7.3.5 Lower in-situ strength

Honeycombing usually means significant loss of strength. Widespread or deep honeycombing must be considered in terms of its effect on the integrity of the structure.

7.3.6 Reliability of investigative technique

So as to establish the significance of a honeycombing or surface damage nonconformity, account must be taken of the reliability of the investigation technique *per se* and of the level of uncertainty associated with testing[3].

7.4 DETERMINATION OF APPROPRIATE REMEDY

Remedial options for honeycombed and surface damaged concrete include:

- take no action
- make good
- relax the specification requirements
- cut out and reinstate
- demolish and rebuild.

These are discussed below.

It is important to agree who will be responsible for any long-term maintenance commitments and, if necessary, to set aside funds to meet the future costs of the works.

7.4.1 Take no action

Where the extent of the honeycombing or surface damage is small in relation to the section; where the area of honeycombing or surface damage is inconspicuous at the appropriate viewing distance, and if strength and cover requirements are satisfied, it may be appropriate to take no action.

7.4.2 Buttering-up

This is unacceptable. Honeycombed concrete cannot be remedied by buttering-up.

7.4.3 Make good

Where the area of honeycombing or surface damage is shallow (say less than 25 mm deep) the concrete can be repaired by making good. Details of acceptable making good techniques are described in the BCA publication *Making good and finishing*[89].

For aesthetic reasons it may be appropriate to apply a finishing coat of mortar or other decorative coating to the concrete after completing the making-good operations. Such coatings will usually need to be regularly maintained and/or periodically renewed.

7.4.4 Cut out and reinstate

Honeycombing or surface damage deeper than 25 mm will usually need to be cut out and the concrete section reinstated. Repair materials chosen for the reinstatement must be compatible with the original concrete and with any structural or fire requirements[65,96]. In selecting the repair material, consideration should be given to:

- strength properties
- other mechanical properties
- durability properties
- adhesion properties
- shrinkage properties
- thermal properties
- ease of application
- cost.

Consideration should also be given to the aesthetic requirements of the structure, since the repair is unlikely to match the surrounding concrete in colour or texture. Applying a decorative coating may result in a more pleasing appearance, but this option usually has long-term cost implications since any decorative coating will probably need to be regularly maintained and/or periodically renewed (see also Sections 3 and 6). Cutting out defective concrete with tools can lead to damage to the reinforcement, so consideration should be given to the careful use of water jetting techniques.

7.4.5 Demolish and rebuild

Widespread or severe honeycombing, or large areas of surface damage can lead to the conclusion that demolition and rebuilding is the only viable option, particularly where strength, aesthetic or durability requirements cannot be satisfied by other means.

7.5 PREVENTING RECURRENCE

7.5.1 Review mix design

The mix design should be reviewed to avoid:

- excessively high water contents
- non-uniform aggregate gradings
- insufficient fine aggregate.

Honeycombing can occur due to loss of workability arising from weather or local conditions. Honeycombing may also be caused by a rapid loss of workability due to the effects of plasticising admixtures reducing with time. Trials may be necessary to assess these effects.

Mix constituents that give rise to low early strength gain can contribute to the incidence of surface damage at early ages (unless the work is protected). Suitable precautions should be taken when striking formwork. A review of the mix design should therefore address factors that can influence the rate of strength gain, ie:

- use of admixtures
- mix temperature.

7.5.2 Review formwork

Honeycombing or surface damage can arise due to inadequacies in formwork. Consideration should be given to:

- the type of formwork, since this can affect the adhesion between the face of the form and the concrete
- provision of adequate support to ensure forms do not move during concreting, giving rise to grout loss
- use of release agents (thinly and uniformly applied) to facilitate striking
- providing profiles that facilitate release
- sealing formwork joints to prevent grout leakages
- preventing moisture loss from timber formwork
- providing insulation to formwork in colder weather.

Advice on the design of formwork and falsework is well documented[91,98].

7.5.3 Review concreting operations

Honeycombed concrete is often the result of poor compaction.

Honeycombing can be prevented by taking more care with batching, placing and compacting operations. Consideration should be given to the following:

- changing the maximum aggregate size
- ensuring mixing is thorough
- avoiding mix segregation when placing concrete in deep sections
- avoiding mix segregation when placing concrete in sections which contain heavily congested reinforcement
- avoiding excessive loss of fines in moving concrete within the forms
- maintaining a steady supply of fresh concrete during placing operations
- placing concrete close to its final position
- ensuring adequate vibration to fully compact the concrete
- positioning of construction joints.

7.5.4 Striking formwork

Surface damage can be caused by striking formwork too early. Formwork striking times may need to be increased in cold weather or if cement composites are incorporated into the mix. Guidance on formwork striking times is given in CIRIA Report 136[97], together with advice on suitable methods for monitoring the early strength gain of concrete in order to determine the appropriate striking time.

Accidental surface damage can occur during formwork removal operations. Care should be taken to prevent:

- damaging newly exposed concrete during dismantling of formwork and temporary supports
- gouging of concrete when levering-off formwork using wrecking bars.

7.5.5 Protect newly exposed concrete

Newly exposed concrete needs to be protected from cold weather.

Surface damage can occur as a result of accidental collisions. Protection measures for newly exposed concrete include:

- alerting site personnel to the presence of newly exposed concrete;
- erecting barriers around the newly exposed concrete.

7.5.6 Review design

The incidence of honeycombing can be reduced by paying attention to:

- avoiding congested reinforcement
- avoiding concrete shapes that make formwork construction, and placing and compaction of concrete, difficult.

7.6 TYPICAL EXAMPLES OF HONEYCOMBING AND SURFACE DAMAGE NONCONFORMITIES ARISING FROM CASE REVIEWS

Example 1

First pour of a bridge abutment was constructed using grade C50 concrete with air entrainment. On striking the formwork, this was found to have extensive small surface voids. The contractor agreed to construct a large test panel. This also showed extensive small surface voids, indicating problems with the use of cohesive high cement content air-entrained concrete. After much discussion, the client accepted that the specification was too onerous and the concrete mix redesigned. Subsequent pours of the concrete were satisfactory.

Example 2

Patches of honeycombing were found on a concrete bridge pier. In several places the reinforcement was exposed. The contractor was instructed to break out and recast the worst-affected areas. Shallower areas of honeycombing were made good by bush hammering to remove the honeycombing and cutting back the defective areas to sound concrete, before applying a cementitious mortar.

Example 3

A heavily reinforced bridge hinge was cast using a superplasticised grade C50 concrete. On stripping the formwork large areas of honeycombing were found on the hinge and it was evident that the concrete had not completely filled the shutter. Subsequent investigations revealed that the readymix supplier had dispensed only 75 per cent of the agreed amount of superplasticiser, resulting in a lower workability concrete. The hinge was deemed unacceptable and demolished. During the reconstruction of the hinge the concrete workability was checked using a flow table. No honeycombing was found on the reconstructed hinge.

Example 4

Extensive honeycombing was found on striking the formwork to basement walls. Total demolition and reconstruction was rejected as an option because there was pressure to complete the project to meet an opening date. Instead, the honeycombed areas were cut out and recast piecemeal. On completion of the repair, the concrete surface looked reasonable and was deemed acceptable. For the first two years of its life the walls appeared to perform adequately, but thereafter leakage began to occur. Twenty years on, the problem had still not been resolved.

Example 5

On stripping the formwork to the base of a thick wall, an area 2 m × 1 m × 0.8 m was found to contain no concrete. Investigations revealed that the concrete had been incorrectly placed and that compaction procedures had not been followed due to lack of supervision. The defective area was repaired by shuttering the area and then pouring a high flowability concrete through a funnel. Increased supervision on subsequent pours prevented recurrence.

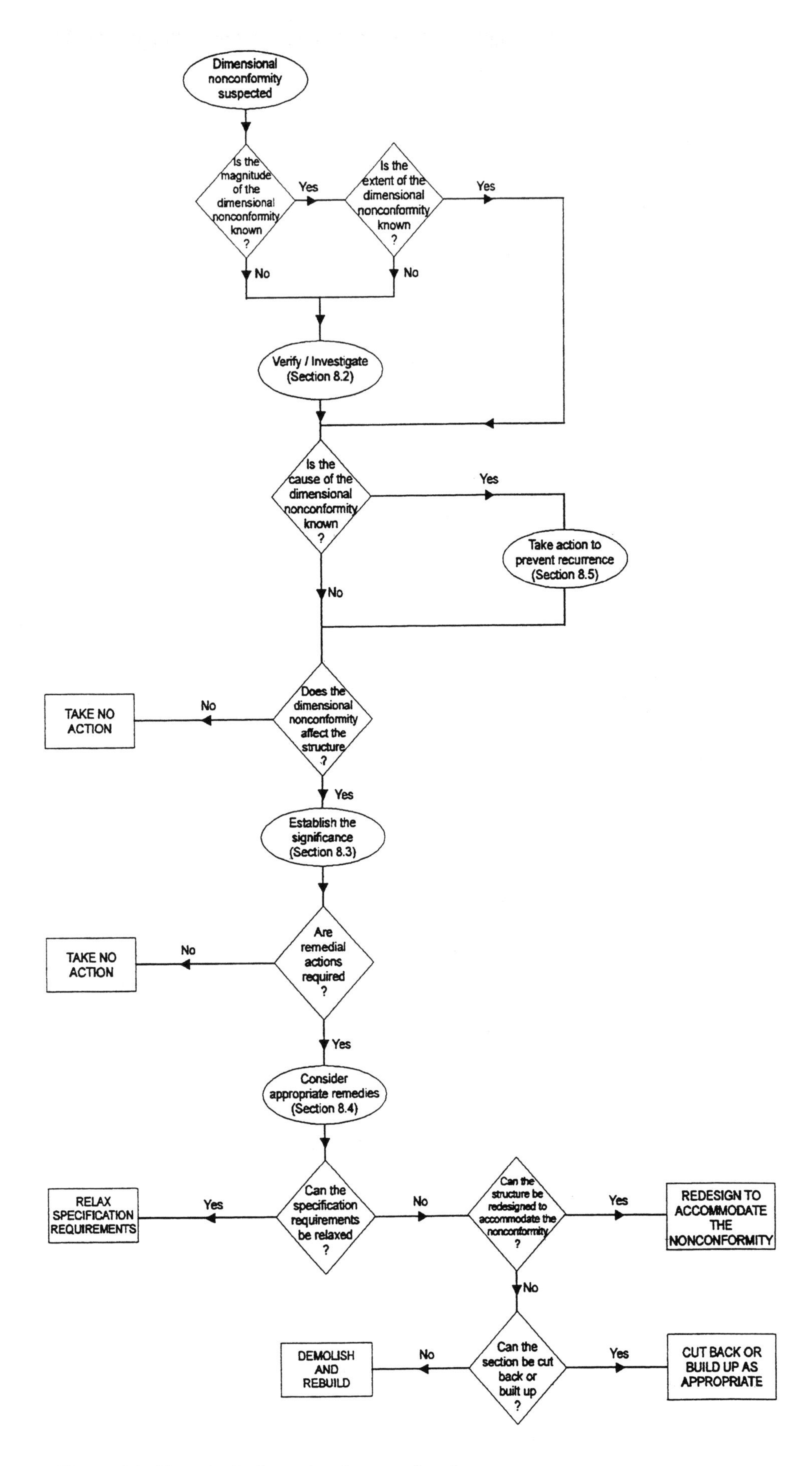

Figure 8.1 *Flow chart: dimensional nonconformity*

8 Dimensional nonconformity

8.1 INTRODUCTION

The decision-making framework for dealing with a dimensional nonconformity is set out in the flow chart in Figure 8.1.

The approach follows the four key processes described in Section 2.

1. Verification/investigation.
2. Establishing the significance.
3. Determination of appropriate remedy.
4. Preventing recurrence.

This section of the report addresses the issues that need to be considered when a dimensional nonconformity is suspected or has been confirmed. Specific dimensional requirements for floors and pavements are not addressed in this report, but guidance can be found in the CIRIA Report *Screeds, floorings and finshes*[126].

In considering the effect of a dimensional nonconformity, reference is made to measurement techniques and to tolerances set down in codes of practice associated with building construction[99-102]. These are also considered to be generally applicable to civil engineering construction. It should be noted that state-of-the-art surveying equipment is capable of achieving better tolerances than those quoted in the current codes of practice, although this does not necessarily mean more accurate construction.

8.2 VERIFICATION/INVESTIGATION

8.2.1 Introduction

Verifying that a dimensional nonconformity has occurred and investigating its extent may involve checks on linear dimensions, height, level, verticality, flatness and skewness. These can be undertaken by taking linear, angular and level measurements. Guidance on the positioning of suitable measuring points when carrying out these measurements is given in BS 7307 [99].

Suitable techniques are listed below and briefly discussed in the next section.

1. Measurement by steel tape (Section 8.2.2).
2. Digital measurement probe/telescopic measuring rod (Section 8.2.3).
3. Plumb line (Section 8.2.4).
4. Inclinometer (Section 8.2.5).
5. Electro-optical distance measurement (Section 8.2.6).
6. Level (optical, digital, laser) (Section 8.2.7).
7. Theodolite (Section 8.2.8).

In carrying out an investigation into a suspected dimensional nonconformity, it is important to select an instrument commensurate with the degree of accuracy required by the specification documents.

8.2.2 Measurement by steel tape

Description of technique

The straightline distance between two points is measured.

Applications and limitations

Steel tapes are commonly used for measuring linear dimensions such as the length, width and thickness of elements and the relative positions of different elements.

Steel tapes are inexpensive, portable and easy to use. Dimensions up to 100 m can be measured. The use of tape tensioners is recommended when measuring distances greater than 10 m[100].

Corrections for temperature, sag and slope may be required[100,103].

Calibration checks against reference tapes should be undertaken regularly. Repaired tapes should not be used unless they have been recalibrated after repair.

Where edges are poorly defined, the use of suitable position pieces can improve measurement accuracy[100].

Reliability

Steel tapes in general use can achieve an accuracy of ±5 mm for measurements of up to 5 m, rising to ±15 mm for measurements of over 25 m. For more precise work an accuracy level of ±3 mm can be achieved for measurements up to and including 10 m[100,103].

8.2.3 Digital measuring probe/telescopic measuring rod

Description of technique

The linear distance between any two surfaces or points is measured by extending the probe/rod to fit the space.

Applications and limitations

The digital measuring probe/telescopic measuring rod is inexpensive, portable and easy to use. It is particularly suited to measuring spaces or openings of up to 5 m in width. Vertical, horizontal and diagonal dimensions can be measured.

The contact surfaces of the probe/rod become worn down with use, so the probe/rod should be regularly calibrated against a known distance.

When measuring vertical dimensions it is important to ensure that the probe/rod is plumb. This can most conveniently be done by using an instrument that incorporates a bullseye level.

Reliability

An accuracy of ±5 mm for measurements of up to 5 m is achievable with a probe/rod instrument.

8.2.4 Plumb line

Description of technique

A vertical reference line is produced using a weight suspended on a wire.

Applications and limitations

Plumb lines are inexpensive, easy to use and portable. They are used to measure deviations from verticality. They can also be used to provide a reference line for offset measurements or when checking variations in surface profile.

The plumb bob mass should be at least 1 kg to ensure that the line remains as stable as possible. Nevertheless air currents can disturb the line, giving rise to errors in readings, particularly when the line is longer than 3 m.

Vibrations in the plumb line can be reduced by immersing the bob in a drum of oil[99,103].

Reliability

The typical accuracy achievable with a plumb line is ±5 m over a 10 m length, provided that the bob is immersed in oil[100].

8.2.5 Inclinometer

Description of technique

Deviations in verticality are measured using the displacement of a spirit level bubble[100].

Applications and limitations

Inclinometers are portable, easy to use and inexpensive. Various types of inclinometer are available.

Measurements along curved surfaces can be made using an inclinometer fitted with studs.

Each recorded reading should be the average of two measurements with the inclinometer reversed between measurements.

Reliability

The accuracy of an inclinometer is dependent upon the sensitivity of the spirit level.

Typical accuracy obtainable with an inclinometer is ±3 mm over distances of less than 2 m.

8.2.6 Electro-optic distance measurement (EDM)

Description of technique

The linear distance between two points is measured by transmitting an electromagnetic signal between the EDM instrument and a reflector[103].

Applications and limitations

EDM instruments are expensive and need to be operated by suitably trained and experienced personnel.

A wide range of EDM instruments is available. EDM signals can be either microwave, infrared or laser. They are capable of measuring distances over longer ranges more rapidly, easily and precisely than steel tapes[103].

Instrument checks should ensure that any frequency drift, index errors or cyclic errors associated with a particular EDM instrument are corrected, or suitable calibration curves produced[103]. Readings are affected by atmospheric conditions since these reduce the velocity of electromagnetic signal. Temperature and pressure corrections may be needed[103].

Reliability

The accuracy of EDM instruments varies with different makes and models.

The accuracy that can be achieved by EDM instruments in general use is ±10 mm for distances over 30 m and up to 50 m[100]. For precise work, using the EDM instrument can typically achieve an accuracy of ±(5 mm + 5 ppm). State-of-the-art EDMs are capable of achieving better accuracies.

8.2.7 Level (optical, digital, laser)

Description of technique

A level comprises a telescope and a spirit level, which are used to establish a horizontal line of sight at each point where the level instrument is set up. Vertical distances above or below the line of sight can be measured using a levelling staff[103].

Applications and limitations

Level instruments vary in price depending on the level of precision required.

Level measurements should be undertaken by experienced personnel. A level is commonly used to determine the elevation of a point above a reference datum, but it can also be used to measure deviations from the horizontal, flatness or skewness.

Several types of level instrument are available[103,104].

Before taking any readings, the collimation error of the instrument should be minimised and parallax eliminated by carrying out appropriate instrument checks.

If possible, backsights and foresights should be of equal length and should not exceed 60 m. Levelling surveys should start and finish at points of known height such as ordnance or temporary benchmarks.

The levelling staff should be placed on a hard surface or footplate and the staff must be held vertical while level readings are recorded. Check that the base of the staff has not become worn with use before beginning levelling operations.

When measuring long distances, curvature and refraction corrections may be required[103].

Reliability

The accuracy achievable by optical level instruments in general use are ±2 mm to ±5 mm over distances up to 60 m[100,104]. Laser levels can typically achieve an accuracy of ±5 m over distances of 100 m. State-of-the-art instruments can achieve great accuracy.

The accuracy of level measurements is affected by the degree of magnification of the instrument telescope, and by the proficiency of the operator. Weather conditions can affect the reliability of the level readings. Errors associated with centring and pointing the instrument can also reduce the reliability of any measurements.

8.2.8 Theodolite

Description of technique

A theodolite is a precision instrument that measures angles in vertical and horizontal planes by sighting onto targets set up at required points[103]. Total stations are instruments combining the functions of a theodolite with EDM.

Application and limitations

Theodolites are expensive, and measurements should be undertaken by experienced personnel. Optical, electronic and laser types of theodolite are available.

Recognised observation procedures should be carefully followed to eliminate collimation and dislevelment effects in the instrument[103]. If possible, the focusing distance should be more than 2 m to avoid errors due to focusing the instrument.

Care should be taken when setting up the theodolite since inaccurate centering of the instrument leads to errors in measuring angles. These errors are larger when sighting over short distances[103].

Reliability

Theodolite measurement accuracy depends on the degree of magnification of the instrument telescope and the proficiency of the operator. The typical accuracy achievable by a theodolite in measuring angles is ±3″ for an instrument capable of measuring to a precision of 1″ [100], although state-of-the-art theodolites are capable of achieving greater accuracy.

Errors associated with centring and pointing the theodolite can reduce the reliability of any measurements.

8.2.9 Other considerations

Investigations of suspected dimensional nonconformity need to consider the influence of the various components in a constructed section. Observed dimensional nonconformity in a particular component may be a consequence of misalignment or dimensional nonconformity associated with another component.

8.3 ESTABLISHING THE SIGNIFICANCE OF DIMENSIONAL NONCONFORMITY

The significance of a dimensional nonconformity needs to be established in relation to:

- structural integrity
- fitness for purpose
- effect on subsequent operations
- durability
- aesthetics.

8.3.1 Structural integrity

The concrete structure must be capable of carrying the design loading. The effect of the dimensional nonconformity on the original design – usually in terms either of an increase in loading or a reduction in load-carrying capacity – needs to be addressed. Deflections associated with loading need to be allowed for when determining tolerances.

8.3.2 Fitness for purpose

The extent to which the dimensional nonconformity affects the intended purpose of the structure or of part of the structure should be evaluated.

8.3.3 Amount of work at risk

Once a dimensional nonconformity has been verified, subsequent construction operations associated with the nonconforming element may have to be suspended. The magnitude of any dimensional nonconformity and its effect on other components in the structure needs to be evaluated, and the significance addressed in the context of the cost of delays to the construction programme.

8.3.4 Capability of joints to accommodate a misfit

It may be possible to accommodate dimensional variations at joints [105]. The design of joints in the vicinity of a nonconformity needs to be evaluated.

8.3.5 Cumulative effects

Minor dimensional nonconformities in individual elements can accumulate when these elements are combined during construction. The possibility of cumulative tolerances requires careful consideration, as does the effect of compounding a dimensional nonconformity. An insignificant nonconformity in one section repeated over several sections can result in a more serious dimensional nonconformity.

8.3.6 Effect on other trades

Manufactured components are usually produced to higher tolerances than are normally achieved in-situ concrete construction. Any mismatch of dimensional tolerances needs to be addressed since this can give rise to difficulties in assembling these components on or within the concrete structure, which may lead to an unsatisfactory performance.

8.3.7 Accuracy levels possible in normal construction

The amount of deviation associated with a dimensional nonconformity should be compared to the range of deviations normally found in construction. Typical acceptable deviations associated with building construction are set out in tables in BS 5606[100]. Alternatively, permitted deviations may have been set down in the specification documents. Deviations associated with the nonconformity falling within these ranges may be deemed acceptable, provided no special accuracy requirements have been specified.

8.3.8 Cover depth

Dimensional nonconformities may result in reduced cover depths to the reinforcement. This should be assessed in accordance with the guidelines set out in Section 3.

8.3.9 Aesthetic considerations

The visual impact of any dimensional nonconformity needs to be addressed in terms of the prestige of the structure, the appropriate viewing distance and the effect of any repairs or adjustments on the visual appearance (see Section 6).

8.3.10 Inherent properties of concrete

Thermal movements, moisture movements and elastic deformation under loading can give rise to dimensional nonconformities during construction as a consequence of the inherent material properties of the concrete. These movements may be reversible, irreversible or cyclic. The likely range of movements should be checked[106-108] and the magnitude of the dimensional deviations and their effect on the structure addressed.

8.3.11 Reliability of investigation technique

In order to establish the significance of a dimensional nonconformity account must be taken of the reliability of the investigation technique *per se* and of the level of uncertainty associated with testing[3].

8.4 DETERMINATION OF APPROPRIATE REMEDY

Once a dimensional nonconformity has been established, the remedial options include:

- take no action if the nonconformity is not significant
- accommodate the nonconformity at joints
- redesign subsequent sections
- relax the specification requirements
- cut back or build up the concrete, as required
- demolish and rebuild.

These options are discussed in turn below.

8.4.1 Take no action

Where the dimensional nonconformity is not significant, where it does not adversely affect subsequent operations, where it does not compromise the structural integrity and where the durability of the structure is not at risk, the dimensional nonconformity may be tolerated and no further action required.

8.4.2 Accommodate nonconformity at joints

It may be possible to accommodate a dimensional nonconformity at joints in the section either by making alterations at existing joints or by the introduction of additional joints. Guidance on the design of joints can be found in BS 6093[109] and CIRIA Report 146[110].

8.4.3 Design review

It may be possible to redesign any unbuilt sections to accommodate a dimensional nonconformity. Before deciding on this option, the costs of the redesign work, possible delays to the construction programme, the costs associated with any modifications and the effect on the appearance of the structure need to be addressed.

Where the designer has specified what prove to be unnecessary accuracy requirements or where the specification leads to conflicting requirements, it might possible to accommodate the dimensional nonconformity by relaxing the specification requirements.

8.4.4 Cut back or build up concrete

It might be necessary to cut back or build up concrete to remedy a dimensional nonconformity.

Cutting back young concrete needs to be undertaken with care as it can result in lower cover depths to reinforcement. It may be necessary to provide extra protection in the form of surface treatments equivalent to the shortfall in cover (see Section 3). To avoid damage to reinforcement, consideration should be given to careful use of water jetting.

Building up concrete may be achieved by applying a render or, for a larger-scale nonconformity, by using sprayed concrete. Checks should be carried out to ensure that the built-up section does not set up unacceptable stresses in the nonconforming element or the structure as a whole. Building up normally affects the appearance of the structure.

Note also that it may not always be possible to enhance the load-carrying capacity of an element by enlarging it, if the element is already under stress.

8.4.5 Demolish and rebuild

Demolition and reconstruction may be the only viable option where large-scale errors occur in the position or alignment of key elements of a structure; or in the case of a dimensional nonconformity which places the stability and/or the long-term durability of the structure at risk; or where aesthetic considerations preclude other remedial options.

8.5 PREVENTING RECURRENCE

8.5.1 Review design

Certain aspects of the design of reinforced concrete structures are known to give rise to dimensional nonconformities. These can often be mitigated by paying attention to detail and to buildability. Consideration should be given to:

- making provision at joints to accommodate dimensional variability
- stating maximum and minimum installed sizes for joints
- making allowances in the design to allow for dimensional adjustments for fixings at critical locations
- avoiding as far as possible using fixings that require preformed holes
- checking drawings, particularly those produced by different parties (eg civil drawings and plant drawings) for compatibility before construction
- using accuracy-check calculations to avoid conflicts between components[105]
- making allowances for conflicts between in-situ construction tolerances and manufactured tolerances
- identifying critical areas and specifying appropriate tolerances for the relevant dimensions
- adopting good defaulting practices[105] to avoid problems of fit
- selecting design details that minimise tolerance constraints
- assessing the combined effects of variability in the separate elements of the construction and making appropriate provision in the design
- minimising requirements for special accuracy
- making clear any specified accuracy requirements
- calculating the range of movements arising from the inherent properties of concrete
- ensuring that the site survey was accurate
- basing design on actual rather than nominal sizes.

8.5.2 Review setting-out operations

Errors in setting out operations inevitably lead to dimensional nonconformities. These can be minimised by paying attention to:

- regularly checking the accuracy of instruments[101]
- accurately positioning grid stations and grid lines
- locating components exactly on their setting out marks.

Guidance on setting-out procedures is provided in BS 5964[102] and the CIRIA publication *Setting-out procedures*[125].

8.5.3 Review site operations

Close supervision of erection procedures by competent and experienced staff ensures that components are correctly placed within specified tolerances. Systematic checking of dimensions should be undertaken at all stages of the construction process. This will ensure that any dimensional nonconformities are detected and rectified early in the construction process. Additional monitoring and control procedures should be considered for critical components and at critical locations.

Ensure that formwork is properly designed and constructed to prevent movement or distortion during concreting.

The proper preparation of formwork surfaces prior to casting sections, preventing formwork movements during concreting operations, and the adoption of appropriate formwork striking times can reduce the incidence of dimensional nonconformities.

8.5.4 Method of construction

The method of construction should be reviewed to ensure that it is not contributing to the incidence of dimensional nonconformities.

8.5.5 Full-scale mock-up

The construction of full scale mock-ups to prove the design and to ensure that adequate provision has been made for dimensional variability at critical locations may be considered for complex structures or sections. Checks can also be made on the allowances for formwork and falsework movement during construction.

8.6 TYPICAL EXAMPLES OF DIMENSIONAL NONCONFORMITIES ARISING FROM CASE REVIEWS

Example 1
Checks on the dimensions of a bridge-deck slab revealed that it was 15 mm thinner than specified, leading to reduced cover depth to the top reinforcement. Checks on the design calculations established that the thinner section was adequate for the design loads. Additional protection to the reinforcement to overcome the shortfall in cover depth was provided by applying a surface coating to the slab surface.

Example 2
Tensioning ducts for an in-situ post-tensioned bridge deck were displaced during concrete casting. The employer wished to demolish the bridge deck, but this course of action would have caused considerable delay and seriously affected the completion of the contract. Assessments carried out by the engineer showed that the existing bridge deck could be accepted, provided that the number of tendons in the ducts was increased. This proposal was adopted and the contract satisfactorily completed on time.

Example 3
The location of a large pile cap was shown on two drawings – a general setting-out plan and a more detailed drawing. The location of the cap was subsequently changed, but only the detailed drawing was amended. The contractor used the setting-out drawing to locate the cap, not noting the discrepancy. The entire cap had to be broken out and reconstructed.

9 Specifying to reduce nonconformities and avoid disputes

9.1 INTRODUCTION

Specifications are a key component in contract documentation. Depending on the particular contract conditions, specifications can be named as contract documents (eg JCT Conditions without Quantities[112], ICE 6th edition[113]), contained within the bill of quantities (eg JCT Conditions with Quantities[114]), or specified as a document in the schedule of contract data (eg NEC Conditions[115]).

Specifications should be reasonable, useable and set out simple, clear and unambiguous requirements. Normally specifications attempt to err on the side of caution to ensure that an adequate structure is built. Depending on specific circumstances, failure to conform to the specification may lead to poorer than intended performance.

Unachievable or ambiguous specifications are a significant factor contributing to the incidence of nonconformities in concrete structures[2,116]. Specifications that make no allowance for the possibility of nonconformities occurring often contribute to the delays and costs that arise when a problem is encountered.

There is no national standard for specification writing for the construction industry in the UK (although guidance on specification writing in general is provided in BS 7373[123]). Commercial model specifications and model trade specifications are available, and many government agencies and similar organisations have their own master specifications[25,76], with which designers and contractors are expected to comply. Advice on writing construction specifications is also available[117], but this advice tends to concentrate on writing style and format and to be less concerned with the technical, product and design information necessary to produce a project specification.

The rest of the section provides outline advice on the writing of specifications that:

- are practicable
- are unambiguous
- avoid subjectivity
- pre-empt disputes and delays arising from nonconformities.

9.2 METHODS OF SPECIFYING

There are four fundamental ways of specifying:

- performance
- method
- example
- personal approval.

Performance specification is appropriate when performance can be defined and measured with sufficient reliability, eg concrete strength. Care is required, however, where the means of achieving a particular performance is not well understood, so that what a contractor has priced to do might not necessarily be what the specifier is expecting to be done.

Method specification is appropriate where the property or performance to be achieved cannot be objectively defined or measured but the means of achieving it can be prescribed and compliance or otherwise with the prescribed method can be effectively monitored.

Specification by example is useful where there would otherwise be a large element of subjectivity in establishing acceptance criteria. For instance, the standard for acceptance of a surface finish may be established by comparison with a trial panel or existing structure, or through the standards of surface finish established nationally by CONSTRUCT panels (see Section 6.5.11).

Specification to personal approval should be avoided unless there is no practicable alternative (discussed in more detail in Section 9.5).

9.3 PRACTICABILITY

There is no simple formula for ensuring that specifications are practicable. The following are useful points for consideration, however.

1. Do not leave detailing or specification writing in the hands of the technically inexperienced, however capable otherwise.
2. Make use of published guidance on practicable detailing and specifying.
3. Where possible, rely on well-tried details and specifications that are known to work. Innovate when warranted, but always with care.
4. Where untried methods or techniques are being contemplated, consider discussing their practicability in advance with those likely to be charged with their implementation.
5. Consider the use of pre-tender mock-ups or site trials of innovative or complex constructions or techniques.
6. For those aspects of a project where the construction method may depend upon the contractor's plant or preferred method of working, or where the contractor may be considerably more familiar with the construction technique than the specifier, consider providing a performance specification and allowing the contractor to put forward a method statement.

9.4 AVOIDING AMBIGUITY

It has been suggested that specifiers are frequently ambiguous because they do not know exactly what they want in advance but they do know what they do not want when they see it. While the veracity of this suggestion is open to debate, there is little doubt that ambiguity is often seen as a means to flexibility. That is, if the specification is not entirely clear at the outset then the specifier has the opportunity to decide what is required as work proceeds.

It is recommended that this approach be avoided under all circumstances, except when the specification makes it clear that it is of itself not definitive on certain issues because those issues are to be resolved during the course of construction. Only in this circumstance can a contractor reasonably be expected to be aware of and to take account of the risks associated with certain specification issues being unresolved at tender stage.

Under all other circumstances, specifications should be clear, unambiguous and incontrovertible. For instance, the approach contained in the New Engineering Contract[115] and the IChemE Conditions[124] is to be strongly encouraged, whereby matters such as requirements for the works, inspection and test requirements, remedial options, etc are set out.

Some specific do's and don'ts are set out below.

Do's

1. Arrange the specification in well-organised sections, so that information can be easily located.
2. Set out acceptance rules, clearly and concisely, including setting out, sampling plans (locations, configuration, etc) and testing regimes, where appropriate.
3. Define the terminology used in the specification and use it consistently throughout the specification, the drawings and any ancillary documentation.
4. Refer to the relevant sections of published standards (BS 5328, BS 8110, etc) whenever possible, using clear and full references and, unless absolutely necessary, do not deviate from or override requirements contained within those standards.
5. State clearly which edition of a standard or standards is incorporated into the specification or use words along the lines of "the relevant editions shall be the latest editions in publication x days prior to the date of tender".
6. Where design standards permit options, make it clear which options are, or are not applicable or, if appropriate, indicate that it is for the contractor to chose.
7. State clearly any parts of a standard that are not applicable.
8. In circumstances where reference is made to other published guides, such as those produced by professional institutions, trade associations or government bodies, quote from the relevant sections of the chosen guide.
9. If appropriate, reformat model trade specifications to suit the specific project.
10. If the contractor is to be responsible for the design of certain elements, make this clear in the specification.
11. Where products are nominated, incorporate manufacturers' instructions pertinent to the use and application of their products in the specification instead of relying on the user reading the small print of product information. (To guard against the possibility of manufacturers changing instructions during the period between specification and actual purchase, it might also be appropriate to specify that those manufacturers latest instructions should be rechecked at the time of purchase).
12. When specifying concrete mixes, use the appropriate schedules from BS 5328[40-43] whenever possible, and complete all of the relevant sections of the schedule properly and unambiguously.
13. In addition to utilising the schedules in BS 5328, where appropriate provide a clear statement of the use of the concrete and set out clearly any pertinent information not covered by the schedules (eg unacceptability of certain aggregate types, etc)[119].
14. Stress to tenderers the importance of reading and understanding the specification for each contract before tendering, and the dangers of assuming that they know what the specification contains because they have undertaken similar work in the past.

Don'ts

1. Do not recycle specifications from one project to another by adopting a cut-and-paste approach, because this invariably creates inconsistencies or results in the inclusion of confusing, conflicting or unnecessary clauses.
2. Avoid long, complex paragraphs; choose words whose meanings are unambiguous.
3. Avoid using pronouns such as "it", "which" and "same". It is better to repeat the relevant noun to prevent misunderstandings.
4. Do not overspecify: if the specification incorporates published standards such as BS 5328 or BS 8110, do not repeat separately in the specification matters that those standards have already dealt with.
5. Avoid subjectivity (see below).

9.5 AVOIDING SUBJECTIVITY

It is often tempting for specifiers, in order to allow maximum flexibility, to use words along the lines of "shall be to the approval of the engineer" (or architect, supervising officer, etc). Wording along these lines may be reasonable when used in relation to matters such as sources of materials, the use of certain types of plant, etc, over which the engineer may wish to permit the contractor flexibility but retain ultimate authority. However, problems often arise where this form of wording is used in respect of a finished product or a parameter. The tendering contractor might have little idea of what is likely to meet the engineer's approval, and might find itself entirely at the whim of the engineer as to what will or will not be approved. Specifications of this type can also place the engineer or architect in an invidious position and can prove difficult to enforce.

One alternative is to specify in entirely objective terms, eg "the minimum value for X shall be Y". This has the advantage of clarity, all parties know exactly what is required, and it will in many cases be the most suitable form of specification.

There are cases, however, where the specifier will wish to establish a required standard but retain some flexibility of action if that standard is not achieved (for instance where a nonconformity has occurred but is not considered to be significant, as discussed elsewhere in this report). It is suggested that in these circumstances a suitable form of words will be along the lines of: "the minimum value for X shall be Y **unless otherwise approved by the engineer**".

9.6 PRE-EMPTING DISPUTES

It is useful to set out before a nonconformity arises the actions to be taken if it does. While it is clearly impossible to anticipate everything that may happen on a construction project, typical nonconformities in respect of cube test results, cover, early-age cracking, honeycombing, surface blemishes, etc can be foreseen and certain actions for dealing with them can be set out in advance. This can be considerably more productive than simply waiting for the problem to occur and then leaving it to the contractor to propose or the engineer to decide, with all of the attendant delays and disputes that can ensue.

Set out below are some issues, discussed in detail elsewhere in this report, which it is suggested can be addressed in specifications, in respect of some of the more commonly occurring nonconformities.

1. What constitutes a nonconformity? (This will be dealt with in most specifications.)
2. How will the occurrence of a nonconformity be verified, ie confirmed that it is a nonconformity and not simply a shortcoming in the sampling and testing regime?
3. Are there different degrees of nonconformity – eg minor, moderate, serious – which warrant different courses of action?
4. In the event of a nonconformity, what further tests are to be carried out?
5. Who is to carry out the extra testing – engineer, contractor, approved third party etc?
6. What is to be the location and frequency of such tests – eg "in the event of a nonconformity, X tests are to be carried out on concrete elements cast on the same day at locations to be selected by the engineer".
7. Who pays for the tests? (This may vary depending on the outcome of the tests).
8. What are to be the acceptance criteria for the test results?
9. What further actions are to be taken if the tests fail?
10. Who is to be responsible for any consequential costs or delays arising from the testing, and does this depend upon the outcome of the tests?
11. If a nonconformity is confirmed, what remedial options will be acceptable, or at least considered, under what circumstances?

It is not proposed that provisions of this type should be incorporated in such a way as to deny the engineer or architect flexibility of action. Properly worded, specifications drafted along these lines can minimise uncertainty on all sides but still allow every situation ultimately to be addressed on its individual merits.

9.7 USE OF EUROPEAN, NATIONAL OR STANDARD SPECIFICATIONS

There are many widely known and recognised European, national and standard specifications presently in use[25,74,76,118,120,121]. While it is not suggested that all of these are perfect, or avoid all of the pitfalls or incorporate all of the suggestions discussed above, they do have the following advantages.

1. They are usually well-known within the relevant industry sectors, ie practitioners are usually familiar with them.
2. They are well tried and many of their inherent "blemishes" have already been identified and ironed out.
3. They are usually subject to regular review and updating.

There can thus be considerable merit in adopting one of these specifications if it is applicable. When adopting a published specification, take into account the following:

- check that there are no constraints on its use (applicability, copyright, etc)
- use the latest version
- those responsible for the design and for the preparation of the contract documentation should read and understand the specification
- avoid making amendments unless absolutely necessary, but insert any necessary project-specific data or clauses.

9.8 MANAGEMENT OF IMPROVEMENT

Specifications that are to be used more than once, either wholly or in part, should be regarded as "living" documents; that is, they should be subject to ongoing review and, where appropriate, developed and refined. Problems encountered and lessons learned on one project should benefit the next; and technological advances, changes to published guidance and relevant standards etc, should be assimilated as soon as is practicable.

In most organisations this is usually best achieved by appointing an individual or group of individuals with specific responsibility for managing the specification or its component parts, by collating and incorporating feedback and by regular review of relevant source material.

9.9 CONCLUSION

The essence of minimising the effects of nonconformity lies in:

- anticipating what may go wrong
- trying to prevent it
- understanding the implications
- having or being able rapidly to formulate a plan of action, preferably set out in the specification, to assess and remedy the situation when a problem occurs.

It is suggested that all of these can be achieved by a combination of careful forethought, clear definition of responsibility and clarity of purpose.

References

1 M LATHAM
Constructing the Team
Final report of the Government/Industry Review of procurement and contractual arrangement in the UK construction industry
HMSO, London, 1994

2 N P KING
Efficient Concrete Practice: A Review of Current Procedure
CONCRETE 2000, Vol 1, pp265–277,
Editors, R K Dhir and M R Jones,
E & FN Spon, London, 1993

3 NAMAS
The Expression of Uncertainties in Testing
NIS 80, 1994

4 J H BUNGEY
The Testing of Concrete in Structures
Surrey University Press, 2nd edn 1989

5 BRITISH STANDARDS INSTITUTION
BS 1881: Part 204: Recommendations on the use of Electromagnetic Covermeters
BSI, 1988

6 J H BUNGEY
Testing concrete in structures: a guide to equipment for testing concrete in structures
CIRIA Technical Note 143, 1992

7 A K KEILLER
The Effect of Bar Spacing on Covermeter Measurements
Cement & Concrete Association, 1985

8 J H BUNGEY, S J MILLARD & M R SHAW
The Influence of Reinforcing Steel on Radar Surveys of Concrete Structures
Construction and Building Materials, Vol 8 No 2, 1994, pp 119–126

9 J BUNGEY
Testing Concrete by Radar
Concrete, Vol 29 No 6, Nov/Dec 1995, pp 28–31

10 BRITISH STANDARDS INSTITUTION
BS 8110: Structural use of Concrete: Part 1, Code of Practice for Design and Construction
BSI, 1997

11 BRITISH STANDARDS INSTITUTION
BS 5400: Design of Concrete Bridges
BSI, 1990

12 BRITISH STANDARDS INSTITUTION
BS 5502: Buildings and Structures for Agriculture
BSI, 1993

13 BRITISH STANDARDS INSTITUTION
BS 6349: Maritime Structures
BSI, 1984

14 BRITISH STANDARDS INSTITUTION
BS 8007: Design of Concrete Structures for Aqueous Liquids
BSI, 1987

15 E J WALLBANK
Performance of Concrete in Bridges: A Survey of 200 Highway Bridges
HMSO, London, April 1989

16 T FEHLHABER & O KROGGEL
Accuracy of Cover Measurement
In: RILEM Workshop on Testing During Concrete Construction, Mainz, March 1990

17 M G K SHAMMAS-TOMAH
The specification and achievement of cover to reinforcement
PhD Thesis, University of Birmingham, 1996

18 L A CLARK, M G K SHAMMAS-TOMAH, D E SEYMOUR, P F PALLETT & B K MARSH
How can we get the cover we need?
The Structural Engineer, Vol 75, No 17, 2 September 1997, pp 289–296

19 M B Leeming & T P O'Brien
Protection of reinforced concrete by surface treatments
CIRIA Technical Note 130, 1987

20 CONCRETE SOCIETY
Surface Treatments for Concrete
Concrete Society Report (in preparation)

21 CONCRETE SOCIETY
Repair of Concrete Damaged by Reinforcement Corrosion
Concrete Society Technical Report No 26, 1984

22 N K EMBERSON & G C MAYS
Significance of property Mismatch on the Patch Repair of Structural Concrete
Mag Conc Res, Vol 42, No 152, pp 147–160, Sept 1990

23 S KOBAYASHI & H KAWANO
Covermeter Development in Japan and Study on Improving Electromagnetic Induction Method
In: RILEM Workshop on Testing During Concrete Construction, Mainz, March 1990, pp 396–403

24 CONCRETE SOCIETY
Spacers for Reinforced Concrete
Concrete Society Report No 101, 1989

25 DEPARTMENT OF TRANSPORT
Specification for Highway Works Series *1700, Structural Concrete*,
HMSO, London, 1991, amended 1993

26 BRITISH STANDARDS INSTITUTION
BS 1881. *Testing Concrete.* Part 101: *Method of sampling fresh concrete on site*
BSI, 1983

27 BRITISH STANDARDS INSTITUTION
BS 1881. *Testing Concrete.* Part 108: *Method for making test cubes from fresh concrete*
BSI, 1983

28 BRITISH STANDARDS INSTITUTION
BS 1881. *Testing Concrete.* Part 111: *Methods of normal curing of test specimens*
BSI, 1983

29 BRITISH STANDARDS INSTITUTION
BS 1881. *Testing Concrete.* Part 116: *Method for determination of compressive strength of concrete cubes*
BSI, 1983

30 J D DEWAR & R ANDERSON
Manual of Ready Mixed Concrete
Blackie, London, 1992

31 BRITISH STANDARDS INSTITUTION
BS 1881. *Testing Concrete.* Part 202: *Recommendations for surface hardness testing by rebound hammer*
BSI, 1986

32 BRITISH STANDARDS INSTITUTION
BS 1881. *Testing Concrete.* Part 203: *Recommendations for measurement of velocity of ultrasonic pulses in concrete*
BSI, 1986

33 BRITISH STANDARDS INSTITUTION
BS 1881. *Testing Concrete.* Part 120: *Methods for determination of the compressive strength of concrete cores*
BSI, 1983

34 BRITISH STANDARDS INSTITUTION
BS 6089 *Assessment of Concrete Strength in Existing Structures*
BSI, 1981

35 CONCRETE SOCIETY
Concrete Core Testing For Strength (including 1987 Addendum)
Concrete Society Technical Report No 11, 1987 (under review)

36 J H BUNGEY
Testing by Penetration Resistance
Concrete, Jan 1981, Vol 15 No 1, pp 30–32

37 BRITISH STANDARDS INSTITUTION
BS 1881. *Testing Concrete.* Part 207: *Recommendations for the assessment of concrete strength by near to surface tests*
BSI, 1992

38 A E LONG & A McMURRAY
The Pull-Off Partially Destructive Test for Concrete In-situ/Non-Destructive Testing of Concrete,
ACI, SP82, 1987, pp 327–350

39 A J CHABOWKSI & D W BRYDEN-SMITH
Internal Fracture Testing of In-situ Concrete
BRE Information Paper IP22/80, 1980

40 BRITISH STANDARDS INSTITUTION
BS 5328. *Concrete.* Part 1: *Guide to specifying concrete*
BSI, 1991

41 BRITISH STANDARDS INSTITUTION
BS 5328: *Concrete.* Part 2: *Methods of specifying concrete mixes*
BSI, 1991

42 BRITISH STANDARDS INSTITUTION
BS 5328: *Concrete.* Part 3: *Specification for procedures to be used in producing and transporting concrete*
BSI, 1991

43 BRITISH STANDARDS INSTITUTION
BS 5328: *Concrete.* Part 4: *Specification for the procedures to be used in sampling, testing and assessing compliance of concrete*
BSI, 1991

44 BRITISH STANDARDS INSTITUTION
BS 1881. *Testing Concrete.* Part 124: *Methods of Analysis of Hardened Concrete*
BSI, 1988

45 CONCRETE SOCIETY
Analysis of Hardended Concrete
Concrete Society Technical Report No 32, 1989

46 W J FRENCH
Concrete Petrography: a review
Quarterly Journal of Engineering Geology, Vol 24, 1991, pp 17–48

47 CONCRETE SOCIETY
Developments in Durability Design & Performance-based Specification of Concrete
Concrete Society Special Publication No 109, 1996

48 RILEM
Performance Criteria for Concrete Durability
RILEM Report 12, J Kropp & HK Hilsdorf (eds)
E & FN Spon, London, 1995

49 CONCRETE SOCIETY
Permeability Testing of Site Concrete: A review of methods and experience
Concrete Society Technical Report No 31, 1988

50 INSTITUTION OF STRUCTURAL ENGINEERS
Appraisal of Existing Structures
Institution of Structural Engineers, 2nd edn 1996

51 B J STUBBINGS, P R AINSWORTH, R CRANE & R A M WATKINS
Appraisal of the Structural Adequacy of Highrise Reinforced Concrete Buildings in Hong Kong
The Structural Engineer, Vol 68, No 16, 21 August 1990, pp 317–326

52 CONCRETE SOCIETY
Use of pfa and ggbs in concrete
Concrete Society Technical Report No 40, 1991

53 B V BROWN
Monitoring Concrete by the Cusum System
Concrete Society Digest No 6, 1984

54 READY MIXED CONCRETE BUREAU
The Essential Ingredient: Testing
A Osborne (ed)
British Cement Association Publication 97.360, 1995

55 J D DEWAR
Increasing the quality of concrete in an existing structure
RILEM Symposium on Quality Control in Concrete Structures, Stockholm, June 1979, Vol 2, pp29–36

56 P R AINSWORTH, R CRANE, G G PAYNE & R A M WATKINS
The strengthening of high rise domestic and industrial buildings in Hong Kong
The Structural Engineer, Vol 75, No 19, 4 October 1994, pp 313–322

57 T H COOKE
Concrete Pumping and Spraying; A Practical Guide
Thomas Telford, London, 1990

58 E S KING
Sprayed Concrete : Properties, Design and Application
S Austin & P Robins (eds)
Whittles Publishing, 1995

59 K D RAITHBY
External Strengthening of Concrete Bridges with Bonded Steel Plates
TRRL Report SR612, 1989

60 A J J CALDER
Exposure Tests on Externally Reinforced Concrete Beams – Performance After Ten Years
TRRL Report 129, 1988

61 R JONES, R N SWARMY & B HOBBS
Bridge Strengthening Using Epoxy Bonded Steel Plates
Highways, Vol 58, No 1963, July 1990, pp 23–25

62 A J J CALDER
Exposure Tests of 3.5 m Externally Reinforced Concrete Beams – The First Eight Years
TRRL Report RR191, 1989

63 FÉDÉRATION INTERNATIONALE DE PRÉCONTRAINTE
Repair and Strengthening of Concrete Structures: Guide to Good Practice
Thomas Telford, London, 1991

64 J L TRINH
Bridge Evaluation, Repair and Rehabilitation in Structural Strengthening by External Prestressing
Editor A S Nowak
Publ NATO/ASI Series E, Applied Sciences Vol 187, 1990, pp 573–582

65 N I C EMBERSON AND G C MAYS
Significance of Property Mismatch on the Patch Repair of Structural Concrete
Mag Conc Res, Vol 42, No 152, pp 147–160

66 J FREARSON
Tests and Testing in Adverse Weather
Concrete, May/June 1995, pp 16–17

67 CONCRETE SOCIETY
Non-structural Cracks in Concrete
Concrete Society Technical Report No 22, 1992

68 D J DANIELS AND C P STRUDWICK
Feasibility study into the detection of macro-cracks in concrete
TRRL Contractors Report 168, 1989

69 CONCRETE SOCIETY
The relevance of cracking in concrete on the corrosion of embedded steel
Concrete Society Technical Report No 44, 1994

70 J BUNGEY
Radar Inspection of Structures
Proc ICE, Vol 99, May 1993, pp 173–186

71 W SUARIS
Crack mode identification in concrete using acoustic emission
Int Conf on NDT, Michigan, June 1993, pp 511–522

72 T A HARRISON
Early-age thermal crack control in concrete
CIRIA Report 91, rev edn 1992

73 C ARYA, J B NEWMAN, L A WOOD
Cracks in Concrete and their Implications for Design (Synopsis)
The Structural Engineer, Vol 72, No 7, 5 April 1994, pp 114

74 BRITISH STANDARDS INSTITUTION
Eurocode 2: *Design of Concrete Structures. Part 1, General rules and Rules for buildings*
BSI, 1992

75 A W BEEBY
Corrosion of reinforcing steel and its relation to cracking
The Structural Engineer, Vol 56A, No 3, March 1978, pp 77–81

76 WATER AUTHORITIES ASSOCIATION
Civil Engineering Specification for the Water Industry
Water Research Centre, 3rd edn 1989

77 C TURTON
A 'Cracking' design
Concrete, Vol 29, No 1, Jan/Feb 1995, pp 29–30

79 C A CLEAR
The effects of autogenous healing upon the leakage of water through concrete
Cement & Concrete Association Technical Report 559, 1985

80 R T A ALLEN & S C EDWARDS (eds)
The Repair of Concrete Structures
Blackie and Sons, London, 1987

81 J C BEECH
The selection and performance of sealants
BRE Information Paper 25/81, Dec 1981

82 BUILDING RESEARCH ESTABLISHMENT
Simple measuring and monitoring of movement in low-rise buildings Part 1: Cracks
BRE Digest 343, April 1989

83 BRITISH CEMENT ASSOCIATION
Plastic Cracking of Concrete
British Cement Association, 2nd edn 1991

84 P B BAMFORTH & W F PRICE
Concreting deep lifts and large volume pours
CIRIA/Thomas Telford Report 135, 1995

85 W MONKS
Appearance Matters 3: The Control of Blemishes in Concrete
British Cement Association (formerly Cement & Concrete Association), 1987

86 W MONKS
Appearance Matters 8: Exposed Aggregate Concrete Finishes
British Cement Association (formerly Cement & Concrete Association), 1987

87 COMITÉ EURO-INTERNATIONAL DU BÉTON
Tolerances on Blemishes of Concrete
CIB Report No 24, 1974

88 W MONKS
Appearance Matters 1: Visual Concrete: Design and Production
British Cement Association (formerly Cement & Concrete Association), 1987

89 BRITISH CEMENT ASSOCIATION
Concrete on Site 8: Making Good and Finishing
British Cement Association Publication, 1994

91 CONCRETE SOCIETY
Formwork – A Guide to Good Practice
Concrete Society, Special Publication CS030, 2nd edn 1995

92 CONCRETE SOCIETY/CIRIA
Controlled Permeability Formwork
(in draft)

93 M N BUSSELL & R CATHER
Care and treatment of steel reinforcement and the protection of starter bars
CIRIA Report 147, 1995

94 J H BUNGEY, S G MILLARD & M R SHAW
The use of subsurface radar for structural assessment of in-situ concrete
ACI, SP128, 1991, pp497–514

95 BRITISH STANDARDS INSTITUTION
BS 1881. *Testing Concrete.* Part 205: *Recommendations for radiography of concrete*
BSI, 1986

96 CONCRETE SOCIETY
Repair of concrete damaged by reinforcement corrosion
Concrete Society Technical Report No 26, 1984

97 T A HARRISON
Formwork striking times – criteria, prediction and methods of assessment
CIRIA Report 136, 1995

98 BRITISH STANDARDS INSTITUTION
BS 5975: *Code of practice for Formwork*
BSI, 1996

99 BRITISH STANDARDS INSTITUTION
BS 7307: *Building tolerances. Measurement of buildings and building products*
BSI, 1990

100 BRITISH STANDARDS INSTITUTION
BS 5606: *Guide to accuracy in building*
BSI, 1990

101 BRITISH STANDARD INSTITUTION
BS 7334, *Part 1: Measuring instruments for building construction*
BSI, 1990

102 BRITISH STANDARD INSTITUTION
BS 5964: *Building setting out and measurement*
BSI, 1990

103 J UREN & W F PRICE
Surveying for engineers – 2nd edition
Macmillan, London, 1992

104 J UREN & W F PRICE
Lasers in construction – the evolution continues
Proc Instn Civ Engrs, Civ Engng, February 1997, Vol 120, pp 27–35

105 CIRIA
A suggested design procedure for accuracy in building
CIRIA Technical Note 113, 1983

106 BUILDING RESEARCH ESTABLISHMENT
Estimation of thermal and moisture movements and stress: Part 1
BRE Digest 227, 1979

107 BUILDING RESEARCH ESTABLISHMENT
Estimation of thermal and moisture movements and stress: Part 2
BRE Digest 228, 1979

108 BUILDING RESEARCH ESTABLISHMENT
Estimation of thermal and moisture movements and stress: Part3
BRE Digest 229, 1979

109 BRITISH STANDARDS INSTITUTION
BS 6093: *Code of practice for design of joints and jointing in building construction*
BSI, 1993

110 M N BUSSELL & R CATHER
Design and construction of joints in concrete structures
CIRIA/Thomas Telford Report 146, 1995

111 CONCRETE SOCIETY
Notes on the use of radar testing on concrete structures
Concrete Society Technical Note (awaiting publication)

112 JOINT CONTRACTS TRIBUNAL
Standard Form of Contract without Quantities
JCT, 1981

113 INSTITUTION OF CIVIL ENGINEERS
Conditions of Contract & Use in Connection with Works of Civil Engineering
Thomas Telford, London, 6th edn 1991

114 JOINT CONTRACTS TRIBUNAL
Standard Form of Contract with Quantities
JCT, 1981

115 INSTITUTION OF CIVIL ENGINEERS
The New Engineering Contract
Thomas Telford, London, 1993

116 D W CHURCHER & S J JOHNSON
The control of quality on construction sites
CIRIA Special Publication 140, 1996

117 P J COX
Writing Specifications for Construction
McGraw-Hill, 1994

118 INSTITUTION OF CIVIL ENGINEERS
Specification for Piling
Thomas Telford, London, 1988

119 D BILLINGTON & R COTTON
Modern Concrete Specifications, Are They Necessary?
In "Concrete 2000", editors R K Dhir and M R Jones,
E & FN Spon, London, 1993

120 PROPERTY SERVICES AGENCY
General Specification
PSA, 3rd edn 1991

121 NATIONAL BUILDING SPECIFICATION
Code of Procedure for Project Specification
NBS, 1991

123 BRITISH STANDARDS INSTITUTION
BS 7373: *Guide to the Preparation of Specifications*
BSI, 1991

124 INSTITUTION OF CHEMICAL ENGINEERS
Model forms of Conditions of Contract for Process Plant, suitable for Reimbursable Contracts (Green Book)
I Chem E, 1992

125 B M SADGROVE
Setting-out procedures
CIRIA/Butterworths Special Publication 145, 2nd edn 1997

126 M J GATFIELD
Screeds, flooring and finishes – selection, construction and maintenance
CIRIA Report 184, 1998

127 B GILLINSON (ed)
Wall technology
CIRIA Special Publication 87, 1992

128 R A JOHNSON, D S LEEK & M P COPE
Water-resisting basements – a guide. Safeguarding new and existing basements against water and dampness
CIRIA/Thomas Telford Report 139, 1995

129 CIRIA
Manual of good practice in sealant application
CIRIA/BASA Special Publication 80, 1991

130 BRITISH STANDARDS INSTITUTION
DD83: Assessment of the composition of fresh concrete
BSI, London, 1983

131 ENV 1992-1-1 Eurocode 2: Design of concrete structures –
Part 1: General rules and rules for buildings